you fix it: Lawn Mowers

Carmine C. Castellano BAe, MBA & Clifford P. Seitz BA, MA, PhD

ARCO
New York

**TO
ANTHONY JOHN CASTELLANO**

Revised Edition, Fifth Printing, 1983

Published by Arco Publishing, Inc.
219 Park Avenue South, New York, N.Y. 10003

Library of Congress Catalog Card Number 75-22743
ISBN 0-668-02705-3

Printed in the United States of America

W/ SEP '85

CONTENTS

SERVICE PROCEDURES

INDEX

Fault symptom number appears in boldface; page number in lightface

ACKNOWLEDGMENT

ARCO Publications gratefully acknowledges the cooperation and interest of numerous mower manufacturers and accessory suppliers for material used in this publication. Specific acknowledgement is made to the following firms whose photographs are used through their courtesy:

AC Spark Plug, Division of General Motors, Flint, Michigan

Ammco Tools Inc., North Chicago, Illinois

AMF, Western Tool Division, Des Moines, Iowa

Belsaw Machinery Co., Kansas City, Missouri

Clinton Engines Corp., Maquoketa, Iowa

Champion Spark Plug Company, Toledo, Ohio

Dresser S-K Tools, Hand Tool Division, Franklin Park, Illinois

Graham-Lee Electronics Inc., Minneapolis, Minnesota

Heli Coil Products, Division of Mitre Corporation, Danbury, Vermont

Homelite, a Division of Textron Inc., Port Chester, New York

King Electronics Co., Cleveland, Ohio

Pennsylvania Refining Co., Ohio Division, Cleveland, Ohio

Simpson Electric Co., Chicago, Illinois

Sun Electric Corp., Chicago, Illinois

Wheel Horse Products Inc., South Bend, Indiana

INTRODUCTION

YOU FIX IT: LAWN MOWERS

Your You Fix It book will make it possible for you to maintain your power lawn mower in good operating condition. This book provides step by step procedures for solving most lawn mower problems and includes:

- malfunction verification
- fault isolation
- maintenance and service procedures
- most maintenance and service checks

You Fix It starts with what you know—a symptom, such as "the engine won't start" and then takes you through a step by step procedure of WHAT TO DO and HOW TO DO IT.

Let's start with an example.

Symptom—"Engine won't start"

Step 1—Look for this symptom in the Malfunction Index which lists all the anticipated fault symptoms. Next to "Engine won't start" is an index number.

Step 2—Go to that index number section in the book and carry out the steps indicated. The first and most important procedure is to verify that this is really the fault. That means setting up specific initial conditions and then attempting a start. Very often a fault seems to exist only because a critical prestart step or condition was omitted or overlooked.

Step 3—If the fault persists (is verified) perform each of the operations indicated until the fault is isolated. You can't miss.

Step 4—Carry out the indicated step by step service or maintenance procedures and then

Step 5—Follow the check procedures to verify that the fault has been corrected and eliminated. In this example, that the engine will start.

You Fix It has been organized so that you won't have to spend more than the necessary time to get the mower back in operation. This has been accomplished by ordering the step to step procedures in accordance with:
- the ease with which a procedure can be performed. Simple procedures, such as checking for fuel, are conducted first even though the probability that they are the cause may be small.
- experience that a given fault is caused more often by one condition than any of a number of other possibilities.

In preparing You Fix It a search was made of the Maintenance literature. You Fix It has tried to include all of the symptoms and maintenance procedures that experience indicated could occur and that could be fixed with the tools and resources available in the ordinary home workshop.

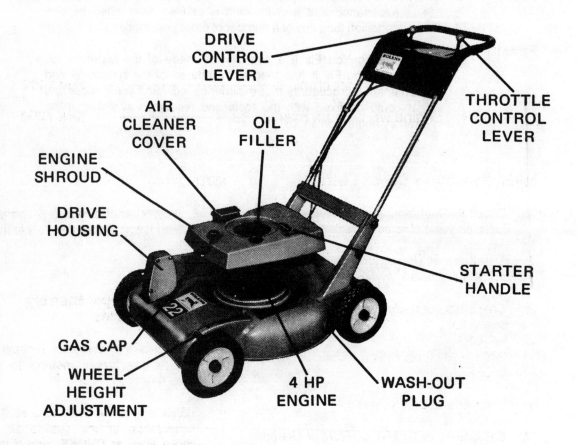

FIGURE 1   Typical Lawn Mower Parts and Controls

ENGINE WILL NOT START

FAULT SYMPTOM 1000

Possible Causes:

- Engine does not crank ——————————— **See 1100**
- Engine will not turn over ——————————— **See 1200**

Initial Conditions Check List:

a. Check the fuel tank. If low, add fuel. Open fuel start valve (if so equipped).

b. Check the oil (for 4-cycle engines). If low, add oil as required.

c. Check that spark-plug cable is connected to spark plug.

d. Set any self propelled mechanism or clutch to OFF or STOP.

e. Set throttle to CHOKE position.

f. Set ignition to START or RUN or ON (for mowers so equipped).

```
———————— CAUTION ————————



        KEEP HANDS, FEET
       CLEAR OF MOWER
       BLADE(S) AND DECK.
        SEE FIGURE 1.



———————— CAUTION ————————
```

NOTE:

The 2-cycle engine requires priming (the adding of a charge of fuel mixture into the engine cylinder).

Priming and Proper Starting of 2-Cycle Engine:

Most engine manufacturers recommend the following as the proper method to start the 2-cycle engine:

- *When engine is cold,* and at the same temperature as the outside air, position choke lever at CHOKE (and if there is a separate THROTTLE, position throttle at middle position of its travel).

- To prime the engine, pull the starter cord out slowly to its full length (on impulse or recoil starters, wind up only part way and slowly release—for electric starters, activate starter with ignition OFF). Engine is now primed.

- When the engine starts, place choke lever at the HALF CHOKE position. Leave at this position for 10 to 20 seconds, then position choke lever to full open position.

- This completes the proper start-up procedure for 2-cycle engines.

1010

Attempt to start the engine

Step 1—Did the engine start?

YES—Fault is not verified. Proceed to Step 3.

NO—Proceed to Step 2.

Step 2—Did the starter crank or turn over the engine?

YES—Proceed to Fault Symptom **2000**.

NO—Proceed to Fault Symptoms **1100** and **1200**.

Step 3—If engine does start, repeat start 4 or 5 times. (Do not repeat priming.)

Did the engine start each time?

YES—Fault has been isolated: An initial condition for start was not performed, or an improper start procedure was used.

NO—Proceed to Fault Symptom **2000**.

ENGINE DOES NOT CRANK
STARTER FAULTS

FAULT SYMPTOM   1100

Possible Causes:

- Electrical connections not properly made ————— **See 1110**

- Jammed starter motor and engine gear ————— **See 1120**

- Battery power, low ————————————— **See 1130**

- Defective starter unit ————————————— **See 1140**

- Defective rope pull cord, manual wind-up ——— **See 1150**

- Defective rope cord rewind spring ————— **See 1160**

- Defective starter mechanism ———————— **See 1170**

Initial Conditions Check List:

a. Perform check list from Fault Symptom **1000**.

b. Check to see that all electrical connections have been properly made on mowers with electric starters.

c. Pay particular attention to cables and wires which run from the mower start-up battery or power package to the electrical starter motor. (See Figs. 2, 3 and 4 for some typical examples of starter motors/cables.)

d. Check the battery power supply. For a power pack be sure the house electrical outlet used to charge the power pack is the proper voltage (usually 115 volts AC) and is working. An ordinary small table lamp can be used to check the electric outlet.

For a wet cell battery, check that the battery is fully charged. Refer to Fault Symptom **1130.**

e. For manual rope cord pull, rope cord rewind, impulse-recoil starter equipped engines proceed directly to Fault Symptom **1150, 1160, 1170.**

1110

Electrical connections

Step 1—After completing the check list for proper electrical connections, attempt start up.

Did engine start?

YES—Fault has been isolated: Improper or insecure electrical connections.

NO—Proceed to Step 2.

Step 2—Did the engine crank?

YES—Fault has not been verified. Repeat Fault Symptom **1000**. If all conditions have been accomplished, then proceed to Fault Symptom **2000**.

NO—Proceed to Fault Symptom **1120**.

1120

Starter motor jam, electrical starter

Step 1—Check as follows to determine if the electric starter motor gears have jammed with the mower drive gear:

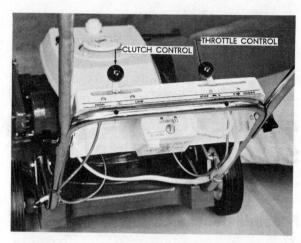

FIGURE 2   Power Pack and Cable Connections

```
————— CAUTION —————

      BEFORE PROCEEDING,
        DISCONNECT THE
       SPARK-PLUG WIRE
      FROM SPARK PLUG

————— CAUTION —————
```

a. Remove mower hood or shroud so that electric motor can be examined.

b. Look at the gear on the electric motor, which engages with the large gear on the engine. The gear on the starter motor should mesh exactly with the large motor gear. There should be no misalignment.

Are the gears jammed?

YES—Fault has been isolated: Jammed starter motor gear. Proceed to Step 2.

NO—Proceed to Fault Symptom **1130**.

Step 2—Break free the jammed gears by turning the top part of engine called the starter cup or starter wheel. See Fig. 5. Rotate the starter cup in both directions, then turn in a COUNTER-CLOCKWISE direction.

Did the gears break free?

FIGURE 3   Typical Electric Starter Mower Cable Run

YES—Fault verified. Proceed to Step 3.

NO—The starter unit must be removed from engine to free jam. Refer to Service Procedure **4042**.

Step 3—Replace hood, connect spark-plug wire and attempt start up.

Did the engine start right up?

YES—Fault isolated, verified and corrected.

NO—Proceed to Fault Symptom **1130**.

1130

Battery, power supply low, electric starter

Step 1—Since the electrical connections are secure and there is no starter motor jam, check that the power supply (the battery or power pack) is sufficiently charged to turn the starter motor, as follows:

a. Lift the hood or shroud so that you can see the starter motor drive gear. See Fig. 5.

b. Look at the starter motor and turn the ignition key ON and OFF a few times.

Did the starter motor operate and crank the engine?

YES—Proceed to Step 2.

NO—The battery power is either low or the starter unit is defective. Proceed to Step 3.

Step 2—Be certain the check list of Fault Symptom **1000** has been followed, then attempt engine start up.

Did the engine start?

YES—Fault has not been verified. Possible oversight in pre-start procedure.

NO—Proceed to Fault Symptom **2000.**

Step 3—The initial conditions check list included check for low battery power. In order to determine the exact condition of the battery or power pack, it will be necessary to measure the charge on the battery. Refer to Service Procedure **4005** and perform the test measurements given.

Did the test measurement of the battery or power pack show a full charge?

YES—Battery charge is good. Proceed to Fault Symptom **1140.**

NO—Fault has been isolated: Low power. Proceed to Step 4.

Step 4—After recharge of battery or power pack, reinstall on mower. Attempt start up of engine.

Did the engine crank?

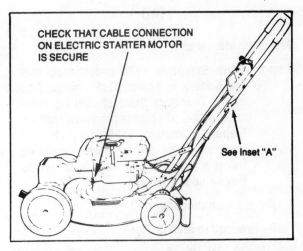

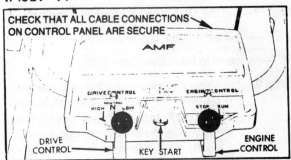

FIGURE 4   Typical Electric Starter Cable Connections

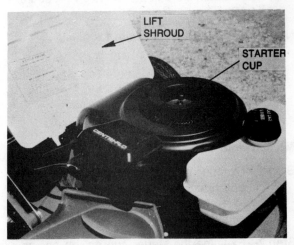

FIGURE 5   Typical Engine Starter Cup

YES—Fault has been verified: Insufficient battery power.

NO—Proceed to Fault Symptom **1140.**

1140

Defective starter unit

Step 1—Fault Symptom **1130** determined that the battery or power pack charge was good. Therefore there should be sufficient power to activate the starter motor. Repeat Symptom **1130**, Step 1, and observe if the starter motor operates using a battery which now is known to have a good charge.

Did the starter motor operate?

YES—Proceed to Step 2.

NO—Proceed to Step 3.

Step 2—Did the starter motor crank the engine?

YES—Proceed to Fault Symptom **2000**.

NO—Proceed to Fault Symptom **1200**.

Step 3—On some engines equipped with electrical starters, there is provision made to start the engine manually using a simple rope pull cord on the starter cup (noted in Fig. 5). Use the rope pull, and attempt to start the engine.

Does the engine crank?

YES—The starter unit may be defective. Remove and check the unit per Service Procedure **4042**.

NO—Proceed to Fault Symptom **1200**.

1150

Manual wind-up, rope pull starter

Step 1—Examine the rope. It is essential that the end of the rope be tied into a knot that slips into the notch on the engine starter cup. See Fig. **5**.

Step 2—Examine the starter cup on the engine and the rope. Both must be clean and free of grease or oil. Grease on the cup or rope will cause the rope to slip.

Is the rope knotted at the end and free of grease?

YES—Proceed to Step 3.

NO—Knot rope. Proceed to Step 3.

Step 3—Clean the rope (or procure a new one since it may be difficult to remove all grease or oil from rope). Clean the starter cup.

Attempt to crank the engine.

Did the engine crank?

YES—Fault verified. If engine does not start, proceed to Fault Symptom **2000**.

NO—Proceed to Fault Symptom **1200**.

1160

Automatic rewind, rope pull starter

Step 1—Does the rope rewind back (recoil) into the starter unit?

YES—Proceed to Step 2.

NO—Fault has been isolated: Defective rewind unit. Proceed to Fault Symptom **2320**.

Step 2—Slowly pull on the rope. Do you feel any resistance?

YES—Proceed to Step 3.

NO—Fault has been isolated: Defective starter or drive mechanism. Proceed to Fault Symptom **2320**.

Step 3—Can the engine be cranked over by pulling slowly on the rope?

YES—Fault has not been verified: Engine does crank. Refer to Malfunction Index for other symptoms and Fault Symptom **2000**.

NO—Fault verified. Proceed to Fault Symptom **1200**.

1170

Impulse-recoil starter

Step 1—Check the starter operation by attempting to wind-up the starter. Proceed as follows:

a. **Do not** set the starter lever or latch to START or WIND-UP.

b.   Simply turn the wind-up handle a few turns.

Is there a clicking-ratchet sound heard?

YES—Proceed to Step 3.

NO—Proceed to Step 2.

Step 2—Is there any resistance felt when the handle is turned?

YES—Proceed to Step 3.

NO—Starter unit is defective. Fault has been isolated. Proceed to Fault Symptom **2330.**

Step 3—Can the starter be wound up and then released?

YES—Proceed to Step 4.

NO—Fault verified: Defective starter. Proceed to Fault Symptom **2330.**

Step 4—Does the engine crank when starter is wound, then released?

YES—Fault not verified. Refer to Fault Symptoms **1000** and **2000.**

NO—Fault isolated and verified: Defective starter. Proceed to Fault Symptom **2330.**

ENGINE DOES NOT CRANK
ENGINE WILL NOT
TURN OVER

FAULT SYMPTOM   1200

Possible Causes:

- Hard matted grass clippings jamming blade(s),

  blade(s) interference ———————————————— **See 1210**

- Mechanical interference, between blade(s)

  and housing ——————————————————— **See 1220**

- Damaged housing ————————————————— **See 1230**

- Defective starter, jammed ——————————— **See 1240**

- Internal damage to engine ————————————— **See 1250**

Initial Conditions Check List:

- Perform check list from Fault Symptom **1000.**

1210
Grass clippings jamming blade

```
————————— CAUTION —————————

        BEFORE PROCEEDING,
         BE SURE IGNITION
        AND THROTTLE LEVER ARE
         SET TO OFF OR STOP.
        DISCONNECT WIRE FROM
          SPARK PLUG. USE
         CAUTION WHEN HAND-
         LING MOWER. KEEP
         HANDS AWAY FROM
            BLADE AREA.

————————— CAUTION —————————
```

Step 1—Perform the following visual inspections:

a. Inspect the mower blade(s) to see if grass or debris is jamming the blade(s).

b. Check if there is mechanical interference with the blade(s) by inspecting:

- **Rotary blade** for deformation, bent or twisted blade, or badly nicked blade.
- **Reel blades** for deformed, bent blades, or blades out of alignment, such that the reel does not have a smooth cutting action, or blades do not move smoothly over the fixed cutting bar.

Is there debris/grass matting jamming blade?

YES—Fault is verified. Proceed to Fault Symptom **1211,** for rotary mowers, or **1212,** for reel mowers.

NO—Proceed to Step 2.

Step 2—Is there evidence of, or actual, mechanical interference with blade(s)?

YES—Proceed to Fault Symptom **1220**.

NO—Proceed to Fault Symptom **1230**.

1211

Rotary mower debris/matted grass blade jam

Step 1—Proceed as follows:
a. Remove blade (refer to Service Procedure **4002**).

b. Clean out all debris from under the deck and frame.

c. Replace blade per Procedure **4002**. Replace plug and ignition wire.

d. Repeat start-up per Fault Symptom **1000/1100**.

Step 2—Did the engine turn over or crank?

YES—Fault has been isolated. Proceed to Step 3.

NO— Proceed to Fault Symptom **1220**.

Step 3—Did the engine start up?

YES—Fault verified: Matted grass jammed blade.

NO—Proceed to Fault Symptom **2000**.

1212

Reel mower debris/matted grass blades jam

Step 1—Proceed as follows:
a. Clean out all debris from blades.

b. Replace plug and ignition wire.

c. Start up per Fault Symptom **1000/1100**.

Step 2—Did the engine turn over or crank?

YES—Fault has been isolated. Proceed to Step 3.

NO—Proceed to Fault Symptom **1220**.

Step 3—Did the engine start up?

YES—Fault verified: Matted grass jammed blades.

NO—Proceed to Fault Symptom **2000**.

1220

Mechanical interference, between blade(s) and housing

```
┌─────────── CAUTION ───────────┐
│                               │
│   BEFORE PROCEEDING           │
│   BE SURE IGNITION            │
│   AND THROTTLE LEVER          │
│   SET TO OFF OR STOP.         │
│   DISCONNECT WIRE             │
│   FROM SPARK PLUG.            │
│   USE CAUTION WHEN            │
│   HANDLING MOWER.             │
│   KEEP HANDS AWAY             │
│   FROM BLADE(S) AREA.         │
│                               │
└─────────── CAUTION ───────────┘
```

When there is mechanical interference with the cutting action of the blade(s), proceed to Fault Symptom **1221 for Rotary Mowers,** or to Fault Symptom **1222 for Reel Mowers.**

1221

Rotary mower—mechanical interference with blade

Step 1—When there is mechanical interference between mower blade and housing or deck due to a damaged blade, replace blade with a new blade. Refer to Service Procedure **4002**.

Step 2—After completion of Step 1, replace plug and ignition wire. Repeat start-up per Fault Symptom **1000/1100**.

Step 3—Did the engine turn over or crank?

YES—Fault has been isolated. Proceed to Step 4.

NO—Proceed to Fault Symptom **1230**.

Step 4—Did the engine start up?

YES—Fault verified: Mechanical interference with blades.

NO—Proceed to Fault Symptom **2000**.

1222

Reel mower—mechanical interference with blades

Step 1—When there is mechanical interference between any of the blades and the fixed cutter bar, then the blade or the bar may be damaged. Re-inspect the reel as follows:

a. Check if any blade is twisted out of line with the other blades.

b. Check if one or more blades is bent or broken such that the blade(s) strike the cutter bar rather than passing smoothly over the bar.

c. Check the cutter bar to see if it has been bent out of shape or misaligned so that it interferes or blocks the action of the reel blades.

Step 2—Is there any evidence of any of the faults noted in Step 1?

YES—Proceed to Step 3.

NO—Proceed to Fault Symptom **1230**.

Step 3—Fault has been isolated. Removal and repair, or replacement, of the reel is required.

NOTE:

Reel removal and repair represents a formidable task for an individual with no prior mechanical repair experience. In such cases it is recommended the mower be taken to the Authorized Service Dealer.

1230

Damaged housing—rotary mower or reel mower

If the blade(s) of the mower has been determined to be in good condition, then examine the mower housing for damage.

```
┌─────── CAUTION ───────┐
│  BEFORE PROCEEDING     │
│  BE SURE IGNITION      │
│  AND THROTTLE LEVER    │
│  SET TO OFF OR STOP.   │
│  DISCONNECT WIRE       │
│  AND REMOVE SPARK      │
│  PLUG. REFER TO        │
│  SERVICE PROCEDURE     │
│  4004. USE CAUTION     │
│  WHEN HANDLING MOWER.  │
└─────── CAUTION ───────┘
```

Step 1—Examine the housing as follows:

a. Look around the frame for bent up pieces, broken off pieces or parts of the frame which have been pushed out of shape.

b. Particularly examine the area within which the blade(s) operates. There should be no bulges or parts of the mower protruding into the path of the blade(s).

Step 2—Is there any evidence of the faults of Step 1?

YES—Proceed to Step 3.

NO—Proceed to Fault Symptom **1240**.

Step 3—Does the damaged housing interfere with the blade(s)?

YES—Fault has been isolated and verified. Proceed to Step 4.

NO—Proceed to Fault Symptom **1240**.

Step 4—Determine if housing damage can be repaired. If in doubt, consult your Authorized Service Dealer for an extent of damage assessment.

FIGURE 6   Typical Take-off/Output Pulley Location

If housing cannot be repaired and engine is in good condition, refer to exploded line drawings for your brand mower, to order a new mower housing. Use the exploded line drawing as a guide to remove the engine from damaged housing and installation on new housing.

1240

Defective starter, jammed

CAUTION

BEFORE PROCEEDING FARTHER, REMOVE THE SPARK PLUG. REFER TO SERVICE PROCEDURE 4004. RELEASE STARTER (IF IT IS WOUND UP) SO THAT STARTER DOES NOT ROTATE ENGINE CRANKSHAFT! IF STARTER WILL NOT RELEASE, IS FROZEN, STAYS WOUND UP (PARTICULARLY RE-COIL–IMPULSE TYPES) THEN REMOVE STARTER AS A COMPLETE AS-SEMBLY. (REFER TO SERVICE PROCEDURE 4040.)

CAUTION

WARNING!

A WOUND UP STARTER IS A SAFETY HAZARD! DO NOT PROCEED UNTIL STARTER IS RELEASED OR IS REMOVED!

Step 1—Proceed as follows:
a.   Disengage any drive pulley or chain, etc.

b.   Then try turning the engine power take-off or output pulley (see Fig. **6**).

c.   Or slowly rotate the mower's blade, USE CAUTION, so as not to cut or catch fingers!

d.   If crankshaft is turning, a popping sound can be heard at spark-plug opening.

Does the crankshaft turn as evidenced by the ability to rotate the mower blade(s)?

YES—Proceed to Step 2.

NO—Proceed to Step 3.

Step 2—Try to rotate the crankshaft in both directions.

Can the crankshaft be freely turned in both directions?

YES—Fault is not a jammed starter. Proceed to engine cranking. Fault Symptom **2300**.

NO—Proceed to Step 3.

Step 3—Remove the starter assembly. Refer to Service Procedure **4040**.

Can crankshaft be rotated now?

YES—Fault has been isolated. Proceed to Step 4.

NO—Proceed to Fault Symptom **1250**.

Step 4—Defective jammed starter unit. Refer to Service Procedure **4040** for repair. After repair or replacement, repeat Start-up of Fault Symptom **1000/1100**.

Did the engine start up?

YES—Fault verified and corrected.

NO—Proceed to Fault Symptom **2000**.

1250

Internal damage to engine

Removal of the starter in Fault Symptom **1240** and the inability of the engine crankshaft to rotate freely indicates the fault is internal in the engine. Engine has internal mechanical jam or a seized piston. This possibility is increased if the start-up of Fault Symptom **1000** check list showed a lack of oil in the engine (for 4-cycle engines). Major engine overhaul and repair is required. Refer to Service Procedure **4100**.

ENGINE IS HARD TO START
FAULT SYMPTOM   2000

Possible Causes:

- Choke valve not fully closing ———————— **See 2010**
- Clogged or dirty carburetor ———————— **See 2020**
- Plugged up fuel tank vent ———————— **See 2030**
- Loose blade (on rotary mower) ———————— **See 2040**
- No fuel flow into engine ———————— **See 2050**
- Poor, or no spark ———————— **See 2060**
- Engine not cranking over fast enough ———————— **See 2070**
- Engine compression not sufficient ———————— **See 2080**
- Air leaks at engine gaskets or seals ———————— **See 2090**

Initial Conditions Check List:

a. Perform all the pre-start checks given in the Check List of Fault Symptom **1000**.

b. On some lawnmowers a safety switch is incorporated into the design of the stone/debris deflector cover, or also the bagging chute. Follow the mower manufacturer's instructions to ensure that the deflector or bagging chute has been properly secured and that the safety stop switch has been properly depressed.

CAUTION

KEEP HANDS, FEET
CLEAR OF MOWER
BLADE(S) BEFORE
STARTING.

CAUTION
---

c. Always start mower on a clean, debris-free surface such as dry pavement. The mower may be difficult to start on uneven heavy grass areas.

2010

Choke valve not fully closing

Step 1—Refer to Fig. **7**. Remove the air strainer/cleaner assembly so that you can see the choke valve (also called choke butterfly). When the mower control is positioned to START, this valve should be closed.

Step 2—Move the mower control back and forth a few times, then position the control to START.

Does the valve fully close?

YES—Proceed to Fault Symptom **2020**.

NO—Proceed to Fault Symptom **2110**.

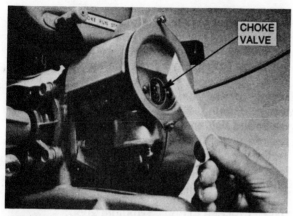

FIGURE 7 Typical Carburetor Choke Valve Location, Air Cleaner Removed

2020

Clogged or dirty carburetor

Step 1—Before replacing the air cleaner assembly, look into the carburetor for evidence of dirt, grass clippings, etc. In order to work properly, the carburetor must be free of all foreign material.

Is there any evidence of dirt or a plugged up carburetor?

YES—Refer to Service Procedure **4010** to properly clean out the carburetor.

NO—Proceed to Fault Symptom **2030**.

2030

Plugged up fuel tank vent

Step 1—Remove the fuel tank cap and examine cap. The cap (on most gasoline mowers) contains a small hole or other openings which vent the fuel tank to the atmosphere. When this vent hole is clogged or closed off the engine will be very difficult to start, since a partial vacuum will develop inside the tank and thereby prevent the fuel from flowing freely to the carburetor.

Is the cap vent clogged?

YES—Proceed to Step 2.

NO—Proceed to Fault Symptom **2040**.

Step 2—Proceed as follows if cap is clogged (or is suspected to be clogged):

a. Shut down the engine if it is running. Place THROTTLE control to OFF. Remove the gas cap.

b. Look at the cap closely and try to see if the small air hole or series of holes in the cap are clogged.

c. On some caps there are no holes as such, but short piece of tubing built into the upper part of the cap. Look into these tubes to see if they have been obstructed by dirt or grass clippings.

d. If the openings appear to be clogged use a sharp pointed tool (a sharp small screwdriver, or a thin hair pin will do) to dislodge the dirt. Obtain a can of carburetor cleaner fluid (at any auto supply shop). Place the gas cap in a small clean can, and pour the cleaner over the cap—covering it completely.

e. Let the cap soak in the cleaner for about 10 minutes

CAUTION

USE MAXIMUM/AMPLE VENTILATION WHEN USING CARBURETOR CLEANER. WORK OUT-DOORS PREFERABLY, AND FOLLOW DIRECTIONS OF CLEANER MANUFAC-TURER—FUMES ARE FLAMMABLE AND HARMFUL

CAUTION
---

f. Remove the cap from the cleaner and shake off the cleaner fluid.

g. Wipe dry. Let the cap dry completely. Blow into the vent openings to assure that the openings are free and clear.

h. When cap is dry replace on tank.

i. Attempt to start up the engine.

Did the engine start right away?

YES—Proceed to Step 3.

NO—Proceed to Step 4.

Step 3—Let engine run. If engine runs well after 5 to 10 minutes or longer without sputtering, the cap has been unclogged. If the engine sputters and slows to a stop, the cap is still clogged.

Does the engine run well, without sputtering:

YES—Fault has been isolated: Plugged air vent in gas cap.

NO—Proceed to Step 4.

Step 4—Place the cap on loosely. Attempt to start up the engine.

Did the engine start right away?

YES—Fault has been verified: Plugged up air vent in gas cap.

NO—Proceed to Fault Symptom **2040**.

2040

Loose blade

NOTE:

The following applies only to rotary mowers which have the blade mounted directly on the end of the crank shaft. For all other mowers proceed to Fault Symptom **2050**.

Step 1—Shut down the engine. Place throttle control to OFF. Remove ignition cable and spark plug per Service Procedure **4004**. Then tilt up the mower and examine the blade.

a. Check the blade to see if it is loose.

b. Move the blade up and down on the crankshaft, and attempt to rotate the blade. USE CAUTION TO AVOID CUTTING YOUR HAND ON BLADE.

Is there any evidence of the blade being loose on the shaft?

YES—A loose blade can cause hard starting. Refer to Service Procedure **4002** for tightening. Then proceed to Step 2.

NO—Proceed to Fault Symptom **2050**.

Step 2—After tightening the blade attempt engine start up. Be sure to reinstall spark plug and connect ignition cable. Attempt start.

Did the engine start right away?

YES—Fault verified: Loose mower blade.

NO—Proceed to Fault Symptom **2050**.

2050

No fuel flow into engine

Step 1—This step will give a simple determination of whether or not fuel is being provided into the engine cylinder. Proceed as follows:

CAUTION

DO NOT PERFORM
THIS TEST ON A
HOT ENGINE—WAIT
UNTIL ENGINE IS
FULLY COOLED TO
THE SURROUNDING
AIR TEMPERATURE

CAUTION
---

a. Disconnect the ignition cable and remove the spark plug per Service Procedure **4004**.

b. Place the THROTTLE control at mid-position or RUN position. If there is a separate choke control, place at CHOKE.

c. Cover the spark-plug opening by holding your thumb lightly over the hole.

d. Crank over the engine several times and remove your thumb.

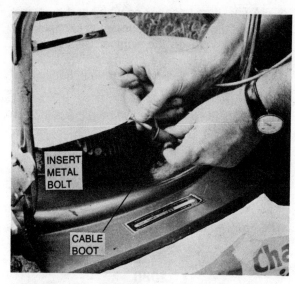

FIGURE 8   Insertion of Metal Bolt into Ignition Cable End Boot

e.   There should be a small amount of fuel on your thumb. The odor of gasoline (or gasoline oil mixture) should be very evident from this small amount of fuel.

Was there a small amount of fuel on your thumb?

YES—Proceed to Fault Symptom **2060**.

NO—Fault has been isolated: No fuel flowing into the engine cylinder. Proceed to Fault Symptom **2100**, and check on fuel system.

2060

Poor, or no spark

Step 1—This step will give a simple determination of whether or not a good spark is being produced by the ignition system. Proceed as follows:

a.   Reinstall spark plug per Service Procedure **4004**.

b.   Place the THROTTLE control at mid-position or RUN position.

c.   If the end of the ignition cable has a protective insulation boot, pull back the boot to expose cap as shown in Fig. 8. If this cannot be done, insert a small metal bolt into the cap as shown in Fig. 8.

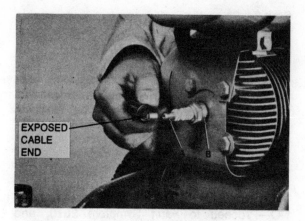

FIGURE 9   Ignition Spark Test

d.   Hold the end of the ignition cable by means of a wooden or plastic clothes pin to avoid the possibility of a porous or leaky cable giving an electric shock.

e.   Now hold the cable so that the end of the bolt, or cable end cap, is about 1/8 inch away from the base of the spark plug, then crank the engine. Be sure engine cranks over at full speed.

f.   A spark which is bright-blue in color should jump from the ignition cable end to the base of the spark plug.

Was there a snappy sharp spark?

YES—Proceed to Step 2.

NO—Refer to Fault Symptom **2210**.

Step 2—Step 1 indicated that the ignition system can deliver a spark. However, the spark plug may be defective. Repeat Step 1, except hold the cable cap or bolt end 1/8 inch away from the tip of the terminal of the spark plug. See Fig. 9. Crank engine.

Was a snappy spark produced?

YES—Since spark was produced, but engine starts hard, check engine start speed. Proceed to Fault Symptom **2070**.

NO—Proceed to Step 3.

Step 3—Since the ignition system was capable of delivering a spark to the spark-plug base but not at the terminal, the spark plug may be defective—in that it is failing when under compression. Refer to Service Procedure **4004,** and remove and replace the spark plug. Replace ignition cable and start up the engine.

Did engine start right up?

YES—Fault has been isolated: Defective spark plug.

NO—Proceed to Fault Symptom **2070,** and check starting speed.

2070

Engine not cranking over fast enough

Step 1—This step will give a simple determination of whether or not the engine is cranking over fast enough to start. Proceed as follows:

a. If the engine has an electric starter motor which is battery operated, refer to Service Procedure **4005** for check of battery having a full charge. A low battery may not supply the energy to crank the engine over fast enough to start.

b. If the engine has an impulse or recoil starter, wind it fully and look at the grass screen on top of engine to determine if the engine spins rapidly when the starter is released. If you can readily see the small screen openings or holes as the screen rotates, the engine is spinning too slowly. If this is the case, or if the starter cannot be fully wound, refer to Fault Symptom **2330.**

c. If the engine has a rope pull type starter, be certain the length of rope is sufficient to fully wrap around the starter drive cup. Use of a rope pull which is too short will not permit the engine to develop sufficient speed to start. If the pull rope is broken off inside the rope wind-up unit, refer to Service Procedure **4040.**

d. A final item to check on whether the engine can be cranked fast enough is to check the friction load on the engine. Disengage all drive pulleys, drive chain and the like—including the self-propelling mechanism or reel. Attempt start up.

Did the engine start right up without these friction loads?

YES—Fault has been isolated: Improper start procedure. Refer to manufacturer's start-up procedures and Fault Symptom **1000** Check List.

NO—Proceed to Fault Symptom **2080.**

2080

Engine compression not sufficient

Step 1—This step gives a quick determination on whether the engine is developing a sufficient compression pressure inside the cylinder. Proceed as follows:

a. *Rope Pull Starters.* For *mowers equipped with a rope pull starter* simply pull on the rope. A definite resistance to the pull should be felt. In the event there is little or no resistance, listen carefully for a hissing sound near the engine head as the engine is cranked. This will indicate possible loose head bolts or a blown out gasket.

Was resistance felt to the rope pull?

YES—Proceed to Fault Symptom **2090.**

NO—Fault has been isolated: Insufficient compression. Proceed to Fault Symptom **2500.**

Is there a hissing sound heard near the engine head as the engine is cranked?

YES—Fault has been isolated: Possible blown head gasket or loose head bolts. Refer to Fault Symptom **2500.**

NO—Proceed to Fault Symptom **2090.**

b. *Impulse or Recoil Starters.* For *impulse or recoil equipped starter* engines, remove the ignition cable and spark plug, and crank the engine. Listen at the plug hole for a popping sound as the engine is cranked. If undetectable, press your thumb over the plug opening and crank the engine. A strong alternate suction and pressure should be felt.

Was there a definite popping sound heard and/or strong suction and pressure felt?

YES—Proceed to Fault Symptom **2090**.

NO—Proceed to Step 2.

c. *Electric Starters. For mowers equipped with electric starters* use the alternate rope pull start and proceed as in Step 1a. If there is no alternate rope pull start, proceed as in the impulse starter equipped engine procedure which is given in Step 1b.

Step 2—If it is suspected that the engine is worn and has had heavy service for a long period of time. Before replacing the spark plug, use a long spout oil can and squirt 6 to 10 drops of motor oil directly into the cylinder, through the spark-plug hole. Clean off the spark plug (per Service Procedure **4004**) before replacing the plug. Replace ignition cable and attempt engine start up. Start up the engine; 3 to 4 tries may be needed.

Did the engine start up?

YES—Fault has been isolated: Engine is in need of mechanical overhaul. Refer to Fault Symptom **2500**.

NO—Proceed to Fault Symptom **2090**.

2090

Air leaks at engine gaskets or seals

Step 1—If the engine indicated that it is develop-ing a compression, shows good ignition, a properly operating fuel system, and is still hard starting, then a check should be made for air leaks at gaskets or seals. Air leaks can destroy the correct air to fuel ratio for good combustion and result in a very lean fuel mixture. This will make hard starting. This would be particularly critical for 2-cycle engines which use the crankcase to pump the fuel-air mixture. A bad crankcase, or gear reduction seal, or a leaking fuel mixture intake valve (reed plate) can also cause hard starting in a 2-cycle engine. For this possibility refer to Fault Symptom **2600. Proceed to Step 2.**

Step 2—Clean off the engine carefully. Remove all grime and dirt, particularly around the engine exhaust and around the carburetor. Examine the metal surfaces which are joined together and have gaskets, particularly at the carburetor attaching surface and the cylinder. If the engine can be started, run the engine at a high speed. Visually check if there is any evidence of fuel bubbling out around any gasket surface. Begin at the carburetor and then to the cylinder head, then to the base of the engine. If there is any indication of fuel bubbling, a gasket is leaking.

Is there any evidence of a fuel leak or fuel bubbling at a gasket?

YES—Fault has been isolated: Engine losing compression from leaky gasket. Refer to Service Procedures **4011, 4060,** or **4070**.

NO—More exhaustive checks for hard starting are required. Proceed to Fault Symptoms **2100, 2200, 2300, 2400, 2500** and **2600**.

ENGINE IS HARD TO START
FUEL SYSTEM PROBLEMS
FAULT SYMPTOM   2100

Possible Causes:

- Choke valve not closing, out of adjustment ——— **See 2110**

- Carburetor throttle lever out of adjustment ——— **See 2120**

- Fouled spark plug ——————————————— **See 2130**

- Clogged air cleaner ——————————————— **See 2140**

- Water in fuel ————————————————— **See 2150**

- Improper fuel mix for 2-cycle engines ——— **See 2160**

- Dirty or improper fuel adjustment settings ——— **See 2170**

- Fuel drips from carburetor, floods engine ——— **See 2180**

- No fuel, defective fuel pump —————————— **See 2190**

Initial Conditions Check List:

Perform all the pre-start checks and conditions given in the Check Lists of Fault Symptom **1000** and **2000**.

```
┌──────── CAUTION ────────┐
│                         │
│     KEEP HANDS, FEET    │
│     CLEAR OF MOWER      │
│     BLADE(S) BEFORE     │
│        STARTING         │
│                         │
└──────── CAUTION ────────┘
```

2110

Choke valve not closing, out of adjustment

Step 1—Perform a visual check as follows to see if the choke valve is completely closing in the carburetor. Refer to Figs. 10 and 11, for typical carburetors and carburetor control cable attachments.

a. Remove the air strainer/cleaner assembly from carburetor so that you can readily see the choke butterfly valve.

b. Move the mower control lever to the CHOKE position and see if the choke butterfly valve moves to a completely closed position.

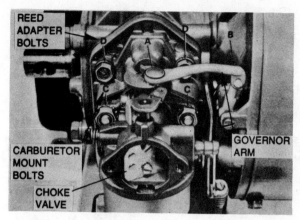

FIGURE 10   Typical 2-Cycle Engine Carburetor

c. If this valve does not fully close, loosen
screws on the choke control cable and
adjust cable until operation of the choke
control fully closes the choke valve in the
carburetor.

d. Tighten screws, replace air cleaner and
start up engine.

Did engine start right up?

YES—Fault has been isolated: Choke valve not
closing properly.

NO—Proceed to Fault Symptom **2120**

2120

Carburetor throttle lever out of adjustment

Step 1—This procedure will determine if the
carburetor throttle lever is opening the
throttle valve far enough. On most
mowers the throttle control also incorporates the CHOKE position, and movement of the control lever to CHOKE
position also positions the throttle valve
to an open position for engine starting.
The following procedure gives an adjustment of a typical throttle control and
carburetor throttle valve:

a. Move the throttle control lever back and
forth and carefully note the position of the
end of the control cable when control lever
is placed at the OFF position.

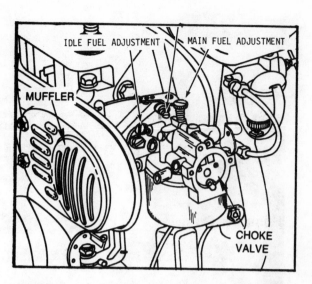

FIGURE 11   Typical 4-Cycle Engine Carburetor

NOTE:

On some mowers it will be necessary to
remove the hood, shroud, or overall engine cover
to observe this and to permit subsequent
adjustments.

b. Move the throttle control lever to the OFF
position. See Fig. 12.

c. Loosen the throttle control cable retaining
screw which is holding the control cable
outer casing in a fixed position. The cable
outer casing should now be free to move.

d. Pull the control cable outer casing in the
direction of the throttle control lever OFF
position—as far as it will go. Do not use a
heavy force so as to kink the cable! Tighten
the cable outer casing retaining screw.

Start up the engine. Did the engine start right up?

YES—Fault has been isolated: Throttle valve
was not opening far enough to permit
start.

NO—Proceed to Fault Symptom **2130**.

2130

Fouled spark plug

Step 1—Crank and activate the starter 4 or 5
times. If engine does not start, remove

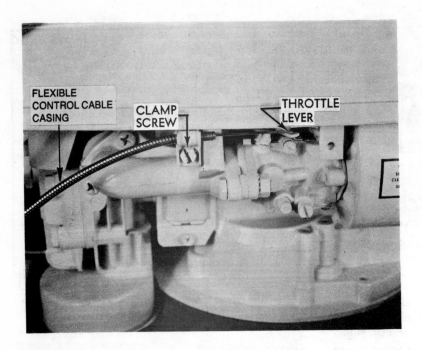

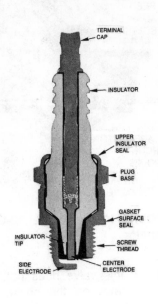

FIGURE 12 Typical Throttle Control Cable Clamping Arrangement

FIGURE 13 Cutaway View of Typical Spark Plug

the spark plug cable and remove spark plug. (Refer to Service Procedure **4004**.) See Fig. 13 for a cutaway view of a typical spark plug and location of electrodes, insulator, etc. Examine the electrodes and insulator tip of the plug. Brown to gray-tan deposits on the insulator show a normal operating, and serviceable plug.

Does the spark plug compare or look like the NORMAL OPERATION plug shown below?

YES—Proceed to Step 2.

NO—Plug is fouled. Refer to Fault Symptom **2210**.

Step 2—If Step 1 showed spark plug to be in good condition, did examination of plug show that the plug was wet with fuel?

YES—Proceed to Fault Symptom **2140**.

NO—No fuel flow, or fuel is not being drawn into engine. Proceed to Fault Symptom **2170**.

2140

Clogged air cleaner

Step 1—Remove the air cleaner filter element and service the filter element as shown in Figs. 15, 16, 17, 18 and 19. Follow the procedure given in Figs. 16, 18 and 19 for adding a light film of oil to the filter element. Clean out the container. Reinstall the filter and replace the air cleaner on the engine, if it was necessary to remove air cleaner. Reinstall the spark plug, connect cable and start up engine.

FIGURE 14 Normal Operation Spark Plug

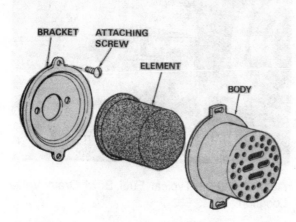

FIGURE 15   Removal of Air Filter

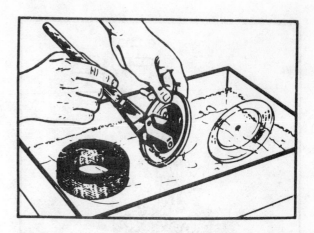

FIGURE 16   Cleaning Metallic Filter

Place the filter in a container of solvent (do NOT use gasoline) and agitate vigorously to remove all dirt and dust from metallic mesh. Dip the air cleaner into oil, then place the filter back on the engine and tighten. Obtain a new filter element if the old one is damaged or becomes lost.

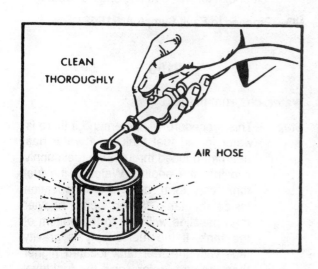

FIGURE 17   Dry Paper Type Filter

("Can" series.) After removal from the engine, brush with a bristle brush (not WIRE); after brushing, use an air hose to blow dirt from the inside to the outside. Do not wet or soak this type filter in solvent or gasoline. Make sure the sealing gasket is in place when reinstalling.

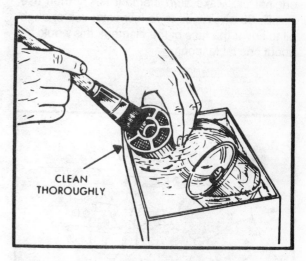

FIGURE 18   Oil Bath Type Air Cleaner

After removal from engine the cleaner should be disassembled and washed thoroughly in a cleaning solvent. If the bowl is made of plastic, make certain that there are no cracks, especially around the sealing areas. If bowl is cracked, replace it. Use compressed air to remove solvent from the mesh filler in the cover. Fill to correct level with S.A.E. 30 engine oil.

FIGURE 19   Polyurethene Element Type Air Cleaner

FIGURE 20   Typical Fuel Bowl Drain Valve Location

After removing the element from container, wash it in hot water, using soap to remove dust, dirt and original oil. Make sure element is dry then use S.A.E. 30 engine oil to re-lubricate; use enough oil to cover the face of the element, this would be about one tablespoon.

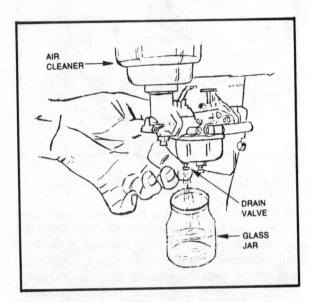

FIGURE 21   Fuel Drain Valve and Glass Jar for Water Test

Did the engine start right up?

YES—Fault has been isolated: Clogged air cleaner filter was causing engine flooding.

NO—Proceed to Fault Symptom **2150.**

2150

Water, dirt, rust in fuel

Step 1—This procedure will determine if there is water in the fuel tank, or if water has been introduced through the fuel supply line into the engine. Water in the fuel tank can form due to condensation inside the tank. Water, being heavier than gasoline, will sink to the bottom of the tank. If the fuel supply is gravity fed—with the fuel tank located higher than the carburetor—or if the fuel level in tank is low, a large amount of water can be drawn into the engine. Even a small amount of water will cause hard starting and if there is any appreciable amount, this will prevent the engine from starting. This condition is most prevalent after a long storage period during which fuel remained in the tank.

Refer to Figs. 20 and 21 for illustration of general fuel tank drain valves, and fuel drain points. Proceed as follows:

a. Open the drain valve and drain some fuel from the tank into a wide-mouth clean transparent glass jar as shown in Fig. 21.

```
┌──────────── CAUTION ────────────┐
│                                 │
│                                 │
│        DO NOT ATTEMPT           │
│        TO DRAIN OR FILL         │
│        FUEL TANK, IF            │
│        ENGINE IS HOT.           │
│        WAIT UNTIL               │
│        ENGINE HAS               │
│        COOLED TO                │
│        PREVENT                  │
│        POSSIBLE FIRE            │
│        IF FUEL SPILLS           │
│        ONTO ENGINE.             │
│                                 │
└──────────── CAUTION ────────────┘
```

b. Fill the jar about 1/3 full, shut off the drain valve and examine the fuel in the jar by holding the jar at eye level. As noted previously the fuel (gasoline or gasoline oil mixture) will float on top of any water and a distinct separation between water and fuel will be evident, if there is water in the fuel.

Is there water in the fuel?

YES—Proceed to Step 2.

NO—Proceed to Fault Symptom **2160** or **2170**.

Step 2—If there is any evidence of water, dirt or rust in the fuel sample, proceed as follows:

a. Completely drain tank into the jar and discard this dangerous fuel.

b. Next, disconnect the fuel line going from the tank into the carburetor and permit this fuel supply line to drain free.

c. Finally, flush out the tank and the supply line with fresh fuel. A final flush using a can of "DRY GAS" (which may be purchased at any gasoline station or auto supply store), is recommended to completely eliminate any residual water in the tank or line.

d. Permit the fuel tank and supply line to dry out ½ to 1 hour.

e. In the meantime dry off the spark plug. Clean and regap the plug per Service Procedure **4004**.

f. Replace the spark plug, connect ignition cable, connect up fuel supply line, close all drain valves and fill tank with clean fresh fuel. Repeat start up.

Did engine start up easily?

YES—Fault has been isolated: Water, dirt or rust in fuel.

NO—Proceed to Fault Symptom **2160**.

2160

Improper fuel mixture, for 2-cycle engines

Step 1—For 2-cycle engines check to determine if the proper gasoline and oil mixture has been used per the engine manufacturer's specification. See Table I for some typical manufacturer's mixtures. If the correct mix is in doubt, drain fuel tank completely, refill with proper mixture, and repeat start up procedure.

Did engine start right up?

YES—Fault has been isolated: Improper fuel (mixture).

NO—Proceed to Fault Symptom **2170**.

2170

Dirty, or improper fuel adjustment settings

Step 1—Activate the starter 3 or 4 times, remove and examine the spark plug. Does the plug show evidence of being wet from fuel?

YES—Proceed to Step 2.

NO—Refer to Service Procedure **4010** for carburetor removal and servicing.

Table I

TYPICAL 2-CYCLE OIL–FUEL MIXTURES

Sleeve Bearing Engine	3/4 pint oil	in	1 gallon gasoline
Needle Bearing Engine	½ pint oil	in	1 gallon gasoline
Jacobsen 321 Engine	1/4 pint oil	in	1 gallon gasoline

USE SAE #30 OR #40 MM RATING OIL. MM IS FOR TYPICAL MODERATE LAWNMOWER SERVICE. USE MS FOR SEVERE OR UNFAVORABLE OPERATING CONDITIONS.

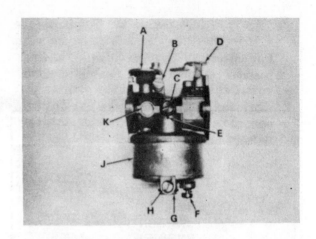

A. **Throttle Plate and Throttle Shaft Assembly**
B. **Idle Speed Regulating Screw**
C. **Idle Fuel Regulating Screw**
D. **Choke Plate and Choke Shaft Assembly**
E. **Atmospheric Vent Hole**
F. **Float Bowl Housing Drain Valve**
G. **Float Bowl Housing Retainer Screw**
H. **High-speed Adjusting Needle**
J. **Float Bowl Housing**
K. **Idle Chamber**

FIGURE 22   Side Draft, Float Bowl Type Carburetor

Step 2—Refer to Figs. **22** to **28** for various makes of carburetors which are typically used on lawnmower engines. Two primary adjustment devices for proper carburetor operation are the main fuel adjustment screw (also called power adjustment needle, or high speed adjustment needle) and the idle fuel flow adjustment screw (also called idle speed adjustment or idle speed adjustment screw, or or idle speed regulating screw). These parts are identified on the carburetors shown in Figs. 22 to 28. Adjust these controls as follows:

The adjustment procedure which follows should be done by hand without any tool: however, it may be necessary to use a screwdriver or wrench to initially loosen the adjustment screws.

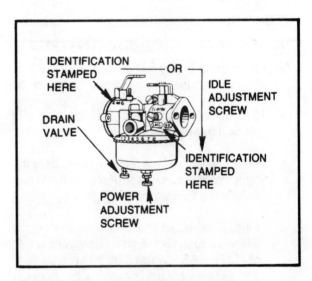

FIGURE 23   Type LMG Carburetor

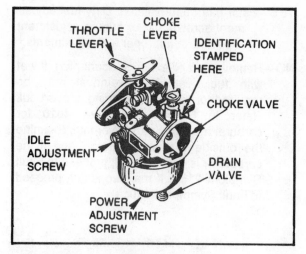

FIGURE 24   Type LMB Carburetor

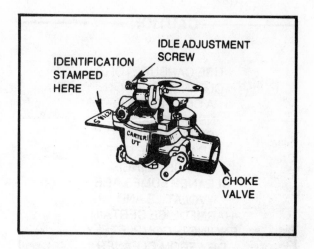

FIGURE 27   Type UT Carburetor

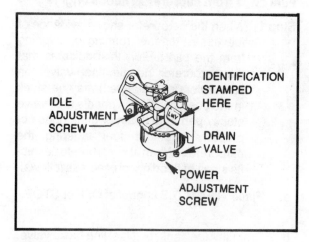

FIGURE 25   Type LMV Carburetor

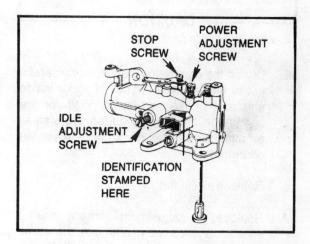

FIGURE 28   Suction Lift Type Carburetor

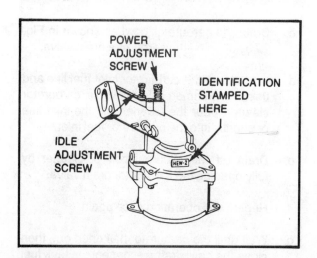

FIGURE 26   Type HEW Carburetor

a. Remove both the main fuel adjustment screw and the idle speed adjustment screw; note carefully from where each screw is removed, and tag each screw if necessary.

b. Place screws in a pan and apply cleaner to thoroughly degrease and clean them. Pay particular attention to the screw tips to insure they are free of all gum and dirt. Dry off the screws.

c. Remove the air filter assembly from the carburetor and apply the cleaner down into the throat of the carburetor, and also in the openings into which the two adjusting screws fit. Apply the cleaner liberally, so that the cleaner runs out of the screw openings.

CAUTION

USE CARBURETOR
CLEANER ONLY ON
A COLD ENGINE.
INSURE MAXIMUM
VENTILATION, PRE-
FERABLY USE OUT-
DOORS—SINCE
CLEANER FUMES ARE
VOLATILE AND
HARMFUL. BE CERTAIN
ENGINE IS COMPLETELY
DRY FROM CLEANER
BEFORE STARTING.

CAUTION

---

d. Dry off the carburetor. Be extremely careful not to leave any lint or bits of paper inside any part of the carburetor. Wait for the carburetor to dry out and then use a small air pump (basketball pump) to blow out the carburetor.

e. Replace air cleaner.

f. Replace the adjustment screws, making sure to replace each one into the same opening from which it was removed. Do not force in these adjusting screws, since they can be damaged easily!

g. Using only your fingers—carefully screw in the main fuel adjustment screw, until it is fully seated and finger tight. After it is seated, back the screw out 1 to 1½ turns. In the same manner, replace the idle fuel adjustment screw until it is fully seated— then back it out 1 turn.

h. Replace the air filter assembly on the carburetor.

i. Dry off the spark plug, reinstall, and attach spark-plug cable.

j. Start up engine.
Did engine start right up?

YES—Fault has been isolated: Dirty fuel adjustment screws, dirt or gum in adjustment lines and/or improper fuel adjustments.

NO—Remove and re-examine spark plug. If wet with fuel, carburetor removal may be required for thorough cleaning or overhaul. Refer to Service Procedure **4010** for carburetor cleaning and overhaul. Examine the outside of the carburetor, and if the carburetor is wet with fuel, proceed to Fault Symptom **2180**. If the plug is dry, proceed to Fault Symptom **2190**.

2180

Fuel drips from carburetor, floods engine

Step 1—When the carburetor shows evidence of fuel leaks, or fuel running or dripping from the carburetor, this indicates that the carburetor needle float valve (on carburetors so equipped) may be stuck in the open position. A needle float valve stuck open, due to dirt or gum, does not properly shut off fuel entering the carburetor. When the carburetor exhibits signs of flooding, proceed as follows:

a. Place THROTTLE control at OFF or STOP.

b. Drain the fuel tank, or if fuel line has a valve, close valve which feeds fuel to the carburetor.

c. Drain the carburetor bowl as shown in Fig. 29 by opening the carburetor drain valve.

d. Disconnect the carburetor inlet fuel line and using a commercially available carburetor cleaner, pour the cleaner into the inlet line or directly into the carburetor fuel inlet.

e. Drain off the accumulation of cleaner by fully opening the carburetor drain valve.

f. Repeat this operation once again.

g. Wait until the carburetor has dried out, then close the bowl drain, reconnect the tank fuel line—open any fuel flow valve(s), and refill the fuel tank, if it was drained.

h. Attempt start up.

Did the engine start right up?

YES—Fault has been isolated: Stuck needle float valve in carburetor.

NO—If the carburetor continues to flood the engine, drip fuel, etc., it is probable that the needle float valve is defective and/or float setting is incorrectly set. In either case the carburetor must be removed for repair and/or proper adjustment. Refer to Service Procedure **4010** for carburetor removal and repair.

2190

No fuel, defective fuel pump

Step 1—On mower engines which are equipped with fuel supply pumps, the pump may be defective. Refer to Fig. **30** for a typical fuel pump installation where the pump is accessible. Engines which do not have an accessible pump require disassembly of the engine. (Refer to Procedure **4020.**) Proceed as follows:

a. Disconnect ignition cable from spark plug.

b. Disconnect fuel supply line running from carburetor to pump. Hold a pan under this open line.

c. Hold the pan at outlet side of pump. Crank over the engine several times.

Does the fuel freely flow out of the pump into the pan?

YES—It has not been verified, that the pump is defective. Proceed to Step 2.

NO—Fault has been verified: Defective fuel pump. Refer to Service Procedure **4020.**

Step 2—If possible, remove the length of fuel line running from pump to carburetor. (If this is not possible use a fine wire to clean out the line.) Immerse the fuel line in carburetor cleaner. Let dry, then blow through line to be sure it's free. Reinstall. Hook up to pump. Reconnect ignition cable and start up engine.

Did the engine start easily?

YES—Fault has been isolated: Clogged fuel supply line.

NO—Refer to Fault Symptom **2000** for other hard-to-start tests.

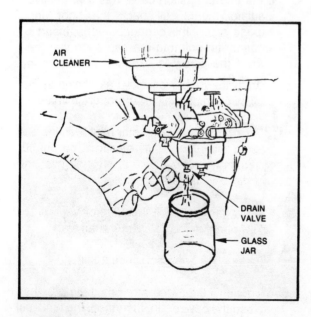

FIGURE 29 Draining Carburetor Fuel into Glass Jar

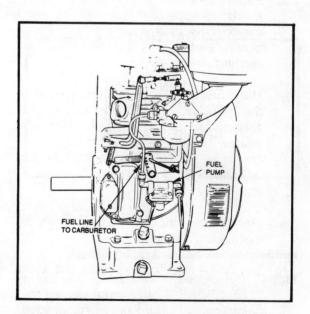

FIGURE 30 Typical Fuel Pump Installation for External Accessible Pump

ENGINE IS HARD TO START IGNITION SYSTEM PROBLEMS

FAULT SYMPTOM 2200

Possible Causes:

- Ignition system not generating a spark ——————— **See 2210**
- Spark plug is fouled ————————————————— **See 2220**
- Spark plug failing in compression ——————— **See 2230**
- Poor spark quality ——————————————————— **See 2240**
- Porous high tension ignition cable ————— **See 2250**
- Defective breaker points ——————————————— **See 2260**

Initial Conditions Check List:

Perform all the pre-start checks and conditions given in the Check Lists of Fault Symptom **1000** and **2000**.

```
┌──────── CAUTION ────────┐
│                          │
│    KEEP HANDS, FEET      │
│    CLEAR OF MOWER        │
│    BLADE(S) AND DECK     │
│    BEFORE STARTING       │
│                          │
└──────── CAUTION ────────┘
```

2210

Ignition system not generating a spark

Step 1—This procedure will establish whether or not a spark is being generated by the ignition system. If Fault Symptom **2060** has been performed, proceed directly to **2220**. If not, proceed as follows:

a. Disconnect ignition cable from spark plug.

b. Place the THROTTLE control at mid-position or RUN position.

c. If the end of ignition cable has a protective insulator boot, pull back the boot and expose cap. If this cannot be done, insert a small metal bolt into the cap such that the end of the screw is exposed as shown in Figure 31.

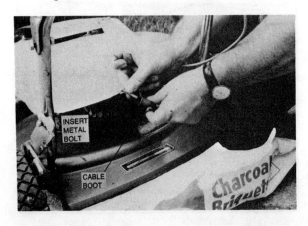

FIGURE 31   Insertion of Metal Bolt into Ignition Cable End Boot

d. Hold the end of the ignition cable by means of a wooden or plastic clothespin to avoid the possibility of a porous or a leaky cable giving an electric shock.

e. Now hold the cable so that the end of the bolt—or the cable end cap—is about 1/8 inch away from the *base* of the spark plug, and simultaneously crank the engine. Be sure engine is cranked over at full speed.

f. A spark which is bright-blue in color should jump from the ignition cable cap, or the bolt end, to the base of the spark plug. The spark will make a snap or crackle sound as it jumps to the plug base.

Was there a snappy bluish-white spark?

YES—Proceed to Fault Symptom **2220.**

NO—Proceed to Step 2.

Step 2—Was there any spark produced?

YES—Proceed to Step 3.

NO—Proceed to Fault Symptom **2250.**

Step 3—Repeat Step 2 several times. Carefully note the color and intensity of the spark produced. If the color was not blue-white and/or the spark did not jump sharply this may indicate a poor ignition/electrical generation system. If the spark color was yellow, orange or yellowish-orange, this represents:

a. A poor spark quality.

b. A spark of insufficient voltage.

Is the spark color any other color than blue-white?

YES—Proceed to Fault Symptom **2240.**

NO—Proceed to Fault Symptom **2230.**

2220

Spark plug is fouled

In the information which follows, most common spark-plug problems are depicted by an accompanying illustration. A visual check of the condition of your spark plug and comparison with the illustration will help isolate why the lawn mow-er is hard to start. In most cases, the spark plug is not worn out and can be cleaned, regapped, and reinstalled.

Remove and examine the spark plug. If the spark plug looks like any of the figures in Steps 1 through 6, the plug is fouled. Refer to Service Procedure **4004** for cleaning and regapping or replace with new plug as noted.

NOTE:

Normal spark-plug service requires removal of the spark plug after 100 hours of mower operation. Check the plug condition, clean and regap as shown in the illustrations of Service Procedure **4004.** Discard a worn out plug and replace with a properly gapped new plug. See Fig. 32 for a normal operating spark plug.

FIGURE 32   Normal Operation Spark Plug

FIGURE 33   Carbon Fouled Spark Plug

Step 1—CARBON FOULED SPARK PLUG— Characterized by fluffy black carbon deposits as shown in Fig. 33. Problems resulting from carbon fouling are:

a. Heavy carbon deposits filling gap between center and side (ground) electrodes short circuit the spark plug. Engine will not start under this condition.

b. Carbon build up from center electrode down side of porcelain insulator to metal base short circuits the spark plug. Engine will not start under this condition.

Causes of carbon fouled spark plug are:

*   Over-rich fuel air mixture. This can in turn be due to dirty or clogged air cleaner which is restricting air flow to the carburetor.
*   Reduced voltage to spark plug. This can result from a poorly operating ignition system, poor contact points, poor magneto, poor condenser, leaky ignition cable. Any of these can cause the plug to misfire and build up carbon deposits.
*   Excessive operation of lawn mower at idle or slow speed. Slow speeds under light load keep the spark plug temperature so low that deposits resulting from normal combustions do not burn off.

To eliminate the causes of carbon fouling, systematically proceed through the Fault Index for each portion of:
*   fuel system.
*   ignition system.
*   mechanical condition.
*   operating conditions.

Step 2—OIL FOULED SPARK PLUG— Characterized by wet oily deposits, insulator tip is blackened, damp oily film over electrodes, and compacting of tarlike gummy carbon around the inside of the plug shell as shown in Fig. 34. Oil fouling is usually accompanied by minimal electrode wear since the plug cannot fire normally. The problem resulting from an oil fouled plug is that the engine will not start or is hard to start depending on the amount of oil build up on plug.

Causes of oil fouled spark plug:

a. A new engine or an engine with newly installed piston rings may sometimes exhibit oil fouling until the engine's piston rings are fully seated. This condition should only be temporary unless the piston rings are defective.

b. In a worn engine oil being pumped by the engine piston past the compression and oil rings, up into the cylinder and building up on spark plug.

c. Engine idle speed set too low.

d. Engine idle adjustment set too rich, or air filter badly clogged.

e. Weak ignition output.

f. Wrong fuel mix of too much oil, wrong type oil.

g. Excessive idling of lawnmower, and not shutting down engine.

h. The valve stem guide (on 4-cycle engines) may have excessive clearance or be worn, thereby permitting oil to be introduced directly into the cylinder. Under this condition the exhaust will be a very thick white smoke with the smell of burned oil—which is exactly what it is.

To eliminate the causes of oil fouling the spark plug, systematically check through the Fault Index on the:
*   fuel system.
*   ignition system.
*   mechanical condition.
*   operating problems.

Step 3—COMBUSTION DEPOSITS FOULED SPARK PLUG—Characterized by brown, yellow and white colored coatings which accumulate on the insulator and ground electrode·as shown in Fig. 35. The problem with such fouling is that the spark plug may not fire at high engine speeds and heavy load causing intermittent missing of the engine.

FIGURE 34   Oil Fouled Spark Plug

FIGURE 35   Combustion Deposits Fouled Spark Plug

FIGURE 36   Overheating and Burned Insulator Fouled Spark Plug

Causes of combustion deposits fouling are:

a.   The combustion process itself wherein fuel and lubrication oil additives produce a powdery by-product.

b.   Excessive combustion chamber deposits.

c.   Clogged exhaust ports or clogged muffler. A heavy coating of these deposits cuts down on the effective area which is available for the spark's operation.

To eliminate the causes of combustion deposits fouling the spark plug, systematically check:
- mechanical condition of engine.
- muffler and exhaust ports.
- proper fuel mixture.

Step 4—OVERHEATING AND BURNED IN- SULATOR FOULED SPARK PLUG— Characterized by dead white blistered coating on the plug insulator tips, burned insulator tips, and badly worn electrodes as shown in Fig. 36. This type of plug fouling can result in hard starting due to the reduction of firing or spark contact surfaces and the increase in the gap size over which the spark must travel. Eventually the fouling builds up to the point where the plug will not fire and the engine will not start.

Causes of overheating and electrode/insulator burning are:

a.   Consistently lean fuel to air ratios.

b.   Dirty or clogged cylinder fins.

c.   Timing over advanced.

d.   Improper installation of the spark plug.

e.   Sustained high speed operation under heavy load which produces high cylinder head temperatures.

f.   Pre-ignition of the fuel air mixture in the cylinder due to combustion of the mixture before the timed spark occurs.

The last item, f, can occur when carbon deposits have built up in the cylinder and such carbon deposits act as hot spots which detonate the fuel air mixture. Except for the pre-ignition problem due to carbon deposit hot spots, most of the causes of this type of plug fouling can be easily remedied by following the lawn mower manufacturer's instructions for manual operation of the mower. Refer to the Fault Index for the Procedure to follow to eliminate fuel system and ignition system problems. A burned insulator spark plug is very difficult to clean and recondition and should be discarded for a new plug.

Step 5—CRACKED, CHIPPED, BROKEN PORCELAIN INSULATOR TIP—Characterized by pieces of the spark-plug insulator being physically broken away or missing as shown in Fig. 37. The engine can be severely damaged if the insulator continues to break apart. Small fragments can score the internal combustion chamber, or jam up and prevent proper operation of valves or piston rings.

Causes of this type of spark-plug failure are heat shock or poor procedures used in gapping or cleaning the spark plug. Heat shock is experienced when there is an excessively fast rise in the insulator tip temperature under severe operating conditions. Heat shock can be caused by improper over-advanced ignition timing which can then cause the fuel air mixture to detonate or ignite spontaneously when the unburned fuel reaches its critical burning temperature.

Remedy of this problem is to carefully follow the gapping procedure shown in Service Procedure **4004.** Refer to Service Procedure **4030** for the procedures to eliminate improper ignition timing, and install a new spark plug, per Service Procedure **4004.**

Step 6—BROKEN OFF CENTER AND/OR GROUND ELECTRODE OF SPARK PLUG—Characterized by a missing portion of spark plug and severely damaged plug shell as shown in Fig. 38. Spark plug in this condition prevents engine from starting and could badly damage the engine. A comprehensive check should be made of all of the engine's systems if the plug is found in this condition.

FIGURE 37   Cracked, Chipped, Broken Porcelain Insulator Tip

FIGURE 38   Broken-off Center and/or Ground Electrode of Spark Plug

FIGURE 39   Incorrect Gasket Seating

FIGURE 40   Correct Gasket Seating

This type of spark-plug failure is caused by overheating, resulting from poor transfer of the heat generated by combustion inside the cylinder. Poor transfer of heat results from poor contact between the spark plug and the engine seat when the spark plug is not properly installed. Improper installation can be caused by dirt at the bottom of the threaded hole or failure to install the gasket. This keeps the plug from being properly seated and the gasket from being properly compressed.

The correct procedure for spark-plug installation is given in Service Procedure **4004**. However, note from Figs. 39 and 40 that the simple procedure of placing the plug gasket in position may easily be overlooked and can cause this problem.

2230

Spark plug failing in compression

Step 1—Reinstall the spark plug and repeat **2210** except do not hold the cable near the plug's base. Hold the cable end cap, or bolt end so it is 1/8 inch away from the tip or terminal of the spark plug. Crank engine as before.

Was there a sharp blue-white spark?

YES—Proceed to Fault Symptom **2500**. Engine may have mechanical compression problems.

NO—Proceed to Step 2.

Step 2—Step 1 **2210** and **2220** comparison showed that a good spark was produced at the spark-plug base, and that the plug appears to be in satisfactory condition by visual inspection. But since no spark was produced at the terminal of the spark plug then the plug may be failing internally when it is subjected to the engine compression. Replace the old plug with a new, properly gapped plug (gap per Service Procedure **4004**). Reconnect ignition cable, follow start-up procedure and attempt start.

Did the engine start up easily?

YES—Fault has been isolated: Defective spark plug. Recheck **2220** to eliminate cause of plug failure.

NO——Proceed to Fault Symptom **2500**.

2240

Poor spark quality

Step 1—The following test will disclose whether or not the basic magneto-electrical system is capable of delivering the required electrical voltage:

a. Disconnect the ignition cable from the spark plug.

b. Obtain a lawn mower spark plug (preferably a new plug, that is known to be good) and open the gap to between 5/32 to 3/16 of an inch as shown in the inset of Fig. 41. (An alternate test plug may be purchased at most lawn mower repair shops. This type of special test plug needs no further adjustments and should be purchased with an attaching clamp. See Fig. 42.)

c. Connect the ignition cable to the terminal of the test plug—and if equipped with an alligator attaching clamp, clamp to an engine bolt or a cooling fin so as to ground the thread portion of the test plug.

NOTE:

Be certain that the engine clamping point is a good ground point. The metal bolt head or fin should be bare and clean—scrape away all paint or dirt before attaching clamp.

d. If the test plug has no attaching clamp use a test lead and clamp one end over the test plug thread and the other end to an engine bolt or fin (be sure to follow the clamping NOTE).

e. Place the THROTTLE control at RUN or at mid-position (turn ignition switch ON if so equipped).

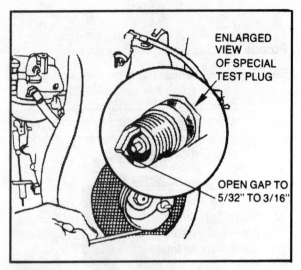

FIGURE 41   Wide Gap Test Plug

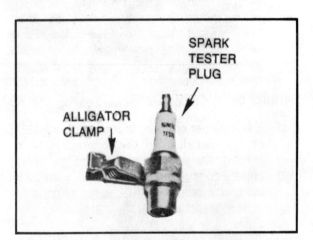

FIGURE 42   Special Test Plug

f. Crank the engine and observe the electrode gap on the test plug. When the engine is properly cranked and with the ignition system in good condition the magneto-electrical system should be capable of delivering more than enough voltage so that a spark will jump across the gap of the test plug (about 10,000 volts).

Was there a spark at the test plug gap?

YES—The ignition system can be assumed to be operating satisfactorily. Proceed to Fault Symptom **2500**.

NO—Refer to Service Procedure **4030** for ignition system repair or overhaul.

2250

Porous high tension ignition cable

Step 1—A fault which must be checked for and corrected, is the possibility of a leaky or defective high tension ignition cable which carries the electricity produced by the engine's ignition system to the spark plug. Proceed as follows:

a. Place controls at OFF or STOP.

b. Begin at spark-plug end of ignition cable—and visually check all around the cable—for its total length, looking for:
- cuts.
- cracks.
- worn spots in the cable insulation.
- where the cable may be worn so thin that the electrical conductor in the inside of the cable can be seen.
- points along the ignition cable which pass over, or come into contact with metal surfaces or corners.

c. If at a corner crossing the cable is worn thin or the electrical conductor actually is in contact with a metal corner, then the cable can be considered to be short circuiting the electricity produced. The result is that no current reaches the end of the ignition cable.

d. If a particular spot is suspect, run inspection at night when a spark can be easily seen.

e. If any spot or point is found on the cable which is as described above, temporarily repair that point by wrapping electrician's tape (available at any hardware store) around that point. The electrician's tape will act as an insulator in preventing the cable from becoming grounded. After this temporary repair connect the cable to the plug, follow start-up procedure and attempt start.

Did the engine start up easily?

YES—Fault has been isolated: Defective high tension ignition cable. Refer to Service Procedure **4030** for permanent repair/replacement of defective cable.

NO—Proceed to Step 2.

Step 2—The high tension ignition cable examination of Step 1 permitted an inspection of the cable up to the point where the cable enters under the engine magneto flywheel or head, but no farther. It is possible the cable may be defective beyond that point and even where the cable is joined to the magneto coil. There is no simple or easy method to examine the cable under the engine magneto flywheel. The engine flywheel must be removed, thereby exposing both the terminal portion of the cable as well as the magneto assembly. The steps for this examination are given in Service Procedure **4030**. Before undertaking this examination proceed to **2260**.

2260

Defective breaker points

NOTE:

This procedure applies to lawn mower engines that are equipped with ignition breaker (also called contact) points which do not require the removal of the magneto flywheel for points examination, or for replacement of the points. Fig. 43 shows the location of such ignition points on a Kohler lawn mower engine. For those engines which do not have accessible points refer to Service Procedure **4030**.

Step 1—The condition of the ignition breaker points are of prime importance for the proper operation of the engine. If the points are burned, oxidized or badly pitted, little or no electrical current will pass through them when they close. The result will be the engine may be hard to start, or will not start at all. In addition, if the engine does start, it will probably run poorly especially at high speeds. Fig. 44 shows some sketches of points in good condition and also points which are badly pitted.

To examine and check accessible breaker points proceed as follows (refer to Fig. 43):

a. Place all controls to OFF or STOP.

b. Remove the breaker points cover and gasket. Handle the gasket carefully so that it can be reinstalled.

c. With a screwdriver, carefully move the end of the movable breaker arm away from the fixed arm so that you can see the flat contact surfaces of the points.

d. If the points are in good condition, they should look like "A" of Fig. 44. If the points are burned or pitted, they will require replacement. Refer to Service Procedure **4030.**

e. After visually examining the points, if they appear to be in good condition, remove the screwdriver, place the THROTTLE control at RUN or at mid-position. (Turn ignition switch ON, if so equipped.)

f. Remove the ignition cable from the spark plug, and while cranking the engine look at the breaker points. If both the points and the ignition system are in good condition a bluish-white spark will jump across the points when they open. This spark should occur consistently. The ignition should give a spark in the same time interval consistently.

Is there a blue-white spark across the points as the engine is cranked?

YES—Proceed to Step 2.

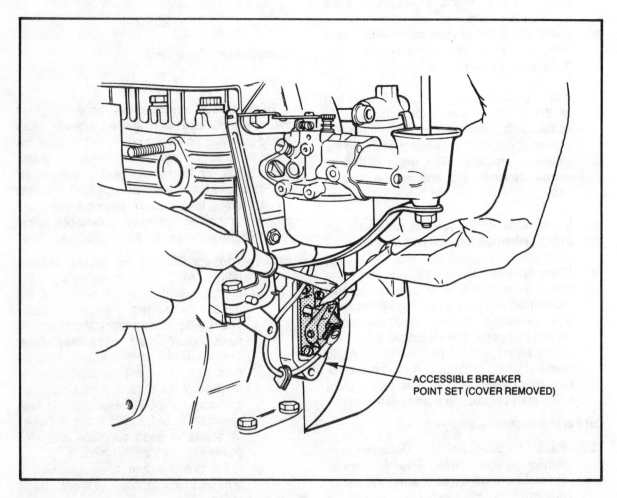

ACCESSIBLE BREAKER
POINT SET (COVER REMOVED)

FIGURE 43   Typical   Accessible   Breaker   Point Set

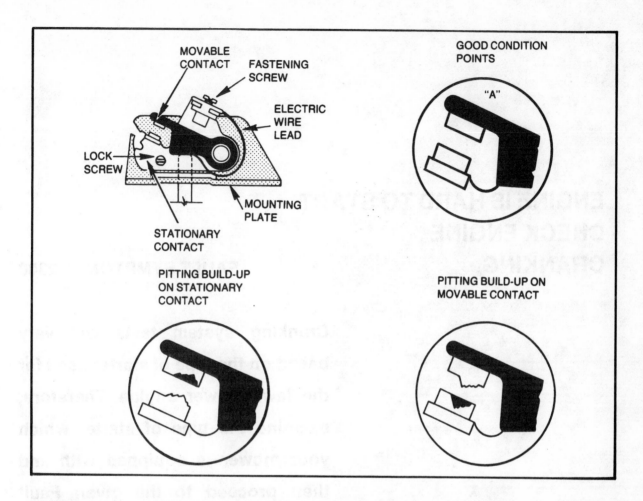

FIGURE 44   Good Condition Breaker Points and Poor Condition Pitted Points

NO—Proceed to Step 3.

Step 2—This test has indicated that the ignition system is capable of producing a good spark. If all the other sections of Fault Symptom **2200** also show that the ignition system is in good condition, proceed to Fault Symptom **2500**.

Step 3—Was there a spark at the points, even if not blue-white?

YES—Proceed to Step 4.

NO—Fault has been isolated: Defective breaker points/defective ignition. Refer to Service Procedure **4030**.

Step 4—Was the spark consistent, produced at even and regular intervals?

YES—Proceed to Step 5.

NO—Fault has been isolated: Defective breaker points/or ignition system component. **Refer to Service Procedure 4030.**

Step 5—Was the spark weak—almost difficult to see?

YES—Fault has been isolated: Poor or defective points/ignition system component. **Refer to Service Procedure 4030.**

NO—It is possible that the ignition system is marginally operational with a spark which does not represent peak output of the magneto system. This could lead to hard starting. Replacement of the breaker points is recommended as a minimal service, then test the system as described in Fault Symptom **2240**. **Refer to Service Procedure 4030** for breaker points service.

ENGINE IS HARD TO START
CHECK ENGINE
CRANKING

FAULT SYMPTOM   2300

Cranking system tests will vary based on the type of starter used for the lawn mower engine. Therefore, examine the type of starter which your mower is equipped with and then proceed to the given Fault Symptom Number cited for your starter type.

a. **Electrical drive starters**—using battery power (or household 110-120 Volt A. C. power), proceed to Fault Symptom **2310**.

b. **Rope pull starters**—Including simple wrap around rope pulls, or with rope recoil, automatic rewind feature, proceed to Fault Symptom **2320**.

c. **Wind-up impulse starters**—proceed to Fault Symptom **2330**.

ENGINE IS HARD TO START
ELECTRICAL DRIVE
STARTER ENGINE

FAULT SYMPTOM   2310

Possible Causes:

- Poor or open electrical cable connections ———— **See 2311**

- Low battery/power pack power ———————— **See 2312**

- Starter unit is defective ———————————— **See 2313**

Initial Conditions Check List:

Perform all the checks given in the Check List of Fault Symptom **1000**.

2311

Poor or open electrical cable connections

Step 1—Place the THROTTLE control to OFF or STOP; turn ignition switch OFF if mower is so equipped.

Step 2—Make a thorough check of all electrical cables. Check for frayed or broken insulation. Check for badly corroded or loose connections. Clean and tighten such connections. If wires are in such a poor condition that repair using electrical insulating tape is too extensive, replace with a new wire. (Refer to the engine manufacturer's exploded view diagram for part number.)

Pay special attention to the wires or cables which go from the mower start-up battery, or power package to the electrical starter motor. (Refer to Fig. 45 for typical example of power-/battery package and cables.)

Did the inspection show any poor or loose connections for the electrical cables?

YES—Proceed to Step 3.

NO—Proceed to Fault Symptom **2312**.

Step 3—With all the cables intact, repaired, or secured, attempt start up of engine.

Did the engine start up easily?

YES—Fault has been isolated: Poor or open electrical cable connection.

NO—Proceed to Fault Symptom **2312**.

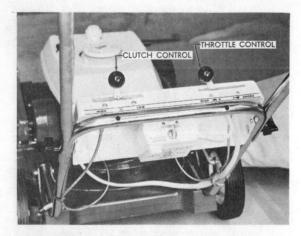

FIGURE 45   Typical Cables and Connections Power Battery Package for Electric Starter Motors

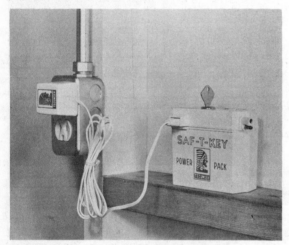

FIGURE 46   Typical Power Package

2312

Low battery/power pack power

This procedure will determine that the battery or power package has a good electrical charge so that it can energize and turn the electric starter motor at full speed.

For sealed power pack type battery proceed to Step 1. For wet cell (lead-acid) type battery, proceed to Step 2.

Step 1—Refer to Fig. 46 which shows a typical dry battery power package. The test procedure for this type of battery is simple:

a.   Make certain the charger unit is plugged into a live outlet and the outlet is of the voltage specified for the charger (usually 115 volt, AC, single phase, 60 cycle electricity).

Check the outlet with a small electric lamp to see if lamp lights.

b.   Make certain the electrical connections between the power package and charger are securely made.

c.   Allow the full time required by the manufacturer, or if the power package is suspected of beginning to fail, for a longer time period, of up to twice the usual time.

d.   If after allowing twice the usual charging time, the power package on the lawn mower fails to activate the starter motor, or the motor cranks the engine very slowly then either the charger unit may be defective, or the battery package is no longer capable of accepting a full charge. Both of these faults may exist together. The most expedient (and economical) way of checking the package's performance, as well as the battery charger, is to take them to an Authorized Service Dealer for your particular brand.

In most cases the service dealer will quickly check—both the package and charger—on a specialized test unit supplied by the factory. Since check out of the package and starter is a very infrequent need, it does not pay to purchase the special test equipment. A power pack should usually last several mower seasons.

e.   Using a power pack, that is known to have a good charge (by either being charged with a good charger unit—or by using a new power pack), reinstall on engine.

Attempt to start the engine.

Did the engine start up easily?

YES—Fault has been isolated: Defective power package and/or charger unit.

NO—Proceed to Fault Symptom **2313**.

Step 2—Wet cell lead-acid type batteries used on some lawn mowers to drive the starter motors can be tested by measuring the voltage level of the battery with a voltmeter. The condition of this type battery can also be measured with a hydrometer. The voltmeter measures the voltage of the battery. The hydrometer tests the specific gravity of the electrolyte in the battery. It is recommended that both a voltmeter and hydrometer be used to measure the condition of the battery.

CAUTION

USE EXTREME CARE
WHEN WORKING WITH
AND AROUND ANY WET
CELL BATTERY AS THE
LIQUID IN BATTERY
IS HIGHLY CORROSIVE
AND CAN CAUSE BAD
BURNS; USE SAFETY
GLASSES TO PROTECT
THE EYES—IMMEDIATE-
LY WASH ANY SPILLED
BATTERY LIQUID OFF
HANDS OR SKIN WITH
WATER

CAUTION

If neither of these instruments are on hand remove the battery by disconnecting the ground terminal clamp first, and then the positive terminal clamp, and have your local gasoline station test the battery.

NOTE:

The positive terminal is the larger terminal, and marked with a (+) and is usually painted red.

Proceed as follows to measure each cell's condition of the battery:

a.   Refer to Figs. 47 and 48 showing the use of a voltmeter to test the battery.

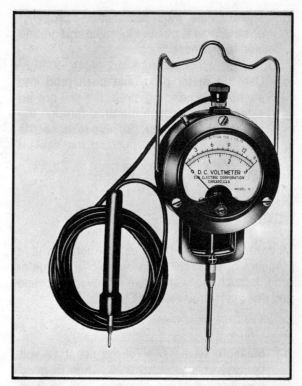

FIGURE 47   Typical Voltmeter for Testing Battery

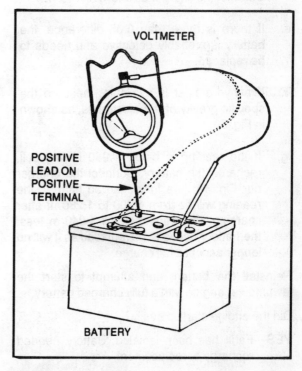

FIGURE 48   Use of Voltmeter in Battery Test

b. The voltmeter must be connected properly, positive lead to positive terminal and ground lead to ground terminal.

c. Use the meter prod, and push prod into each cell connecting strap. If there are no external straps, push the prod through the soft top of the battery. (Be sure to reseal the top after the test by pushing the sealant back together.)

NOTE:

Modern hard case batteries do not permit a measure of each cell by insertion of a voltmeter prod. In such cases measure the overall voltage and use the hydrometer per Step f.

d. Measure each cell's (of the usual 12 volt battery) voltage. Each cell, when in good condition, supplies between 1.95 to 2.08 volts. If there is less than a 0.05 volt difference between the highest and lowest cells, the battery should be recharged. Connect the charger as shown in Fig. 49.

e. If there is more than 0.05 difference, the battery is probably defective and needs to be replaced.

f. If a hydrometer is available, measure the specific gravity of the electrolyte, as shown in Fig. 50.

g. If the reading is below 1.240 in any cell, recharge the battery. Connect the charger per Fig. 49. In a fully charged battery, the reading will be from 1.260 to 1.280. If after charging, the reading is still 1.240 or less, the battery needs to be replaced as it will no longer accept a full charge.

Reinstall the battery and attempt to start the lawnmower engine with a fully charged battery.

Did the engine start up easily?

YES—Fault has been isolated: Battery needed recharge or replacement.

NO—Proceed to Fault Symptom **2313**.

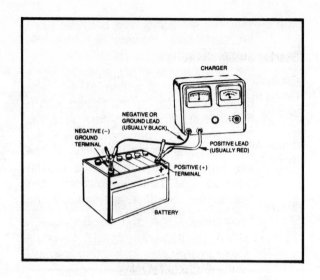

FIGURE 49   Battery Charger Connections

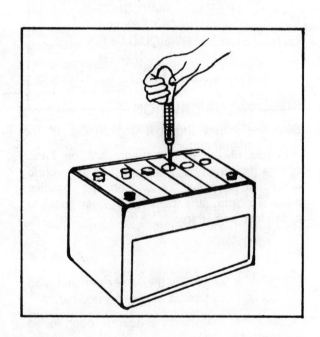

FIGURE 50   Measurement of Battery Specific Gravity with Hydrometer

2313

Starter unit is defective

Fault Symptoms **2311,** and **2312** have established that the power cables and connections are secure and that the battery/power pack has a good charge. In this procedure the electric starter motor and the starter switch will be checked for defects.

Step 1—For the following test it may be necessary to remove the engine cover or shroud so that you can see the starter motor in operation. Check the operation of the motor as follows:

a. Energize the starter mower by briefly opening and closing the starter switch. (Do not hold the starter switch closed!)

b. While the starter switch is ON, look at the starter motor—to see if the drive gear turns.

Does the starter gear turn?

YES—Proceed to Step 2.

NO—Proceed to Step 3.

Step 2—Does the starter motor barely turn?

YES—Fault has been isolated: Defective starter motor. Refer to Service Procedure **4042.**

NO—Proceed to Step 3.

Step 3—Perform the following checks of the starter motor:

a. Check the wiring on the connector of the starter motor.

b. Check for a loose connection, rust or corrosion.

c. Clean and tighten this connection.

d. Check the entire length of the wire for any breaks or cuts, from this connection back to the battery and/or start switch.

e. If no problems are encountered, carefully inspect the start switch. The switch will usually have a single wire going down to the starter motor. This will be the live lead. To complete the connection, part of the switch will be connected to ground.

Carefully check the switch unit. If the switch unit is sealed and cannot be inspected, then the switch can be bypassed with a jumper cable. A jumper cable is an ordinary insulated wire which has alligator clips at each end. This jumper cable will act as the switch. Connect one end to the output wire side of the switch and the other end to a good ground, such as a bolt head on the engine.

NOTE:

Steps a, b, c, d are necessary preliminaries to the successful performance of Step e.

Having bypassed the switch attempt to start engine, does the starter motor engage?

YES—Proceed to Step 4.

NO—The fault has been isolated: Defective starter motor. Refer to Fault Symptom **2323,** or Service Procedure **4042.**

Step 4—Place the engine controls to RUN, ON, or START. With the starter switch bypassed with the jumper cable, attempt start up.

Did the starter motor energize and start the engine easily?

YES—Fault has been isolated: Defective starter switch. Refer to manufacturer's exploded parts diagram for switch part number and replace with a new switch.

NO—Proceed to Fault Symptom **2500.**

ENGINE IS HARD TO START
ROPE PULL STARTER
ENGINE

FAULT SYMPTOM   2320

Possible Causes:

- Rope pull is too short ———————————— **See 2321**

- Starter unit not secure on engine ——————— **See 2322**

- Defects within starter drive or components ——— **See 2323**

Initial Conditions Check List:

a. Perform all the checks given in the Check List of Fault Symptom **1000**.
b. Place THROTTLE control to OFF or STOP. Turn ignition switch OFF, if mower is so equipped.
c. Note the following start up procedures for rope pull starters:

1. Whenever the starter rope is pulled to start engine, never permit the rope to snap back by letting go of the handle. This can dislodge the fixed end of the rope, or break off the rewind spring. Always bring the rope handle back slowly, permitting the recoil spring to uncoil slowly.

2. The rope starter should be pulled in a smooth and steady pull. Do not jerk or yank on the rope, or attempt to pull the rope out as far as it will go.

3. Make a practice of pulling the rope straight out so that unnecessary rope contact and wear is avoided from the rope rubbing against any part of the starter rope guide or hole.

CAUTION

WHEN WORKING WITH REWIND OR RECOIL ROPE STARTERS BE EXTREMELY CAREFUL IN HANDLING THE REWIND SPRING AND THE ROPE PULL AS-SEMBLY. WEAR SAFE-TY GLASSES. IF AS-SEMBLY IS REMOVED FROM MOWER, USE CAUTION, AS THE RE-WIND SPRING MAY ERUPT OUT OF ITS HOUSING IF NOT HANDLED PROPERLY

CAUTION

2321

Rope pull too short

Step 1—Since the engine is hard to start, it is possible that the engine is not cranking fast enough to generate a sufficient voltage to fire the spark plug. Make the following checks:

a. Examine the rope. Check to see that the rope fully retracts into the starter assembly. (Usually the rope should retract right up to the rope pull handle and will be held there by a slight spring tension.)

b. Check if the rope is frayed and stretches when pulled.

If the answer to all of these checks is:

YES—Proceed to Step 2.

NO—Proceed to Fault Symptom **2322** for check on starter rewind.

Step 2—Is the rope very tightly wound in the rewind unit so that you must pull on the rope extra hard?

YES—It is very probable that the rope is too short. A short rope will not crank the engine a sufficient number of revolutions to reach starting speed. Refer to Service Procedure **4040,** for rope pull servicing. Proceed to Step 3 after servicing rope pull.

NO—Proceed to Fault Symptom **2322.**

Step 3—Attempt to start engine.

Did engine start?

YES—Fault corrected. Defective rope pull.

NO—Check on internal starter defects in Fault Symptom **2323.**

2322

Starter unit not secure on engine

Step 1—Check the starter assembly mounting screws or bolts. They must be secure and tight. If they have loosened due to vibration, then there can be undesirable movement or play between the starter assembly and the engine crankshaft. This can result in hard starting.

Are the starter assembly mounting fasteners tight and secure?

YES—Proceed to Fault Symptom **2323.**

NO—Fault has been isolated: Loose mounting bolts or screws on starter unit. Tighten fasteners.
Proceed to Step 2.

Step 2—Attempt engine start.

Did engine start?

YES—Fault corrected. Loose starter.

NO—Proceed to Fault Symptom **2323.**

2323

Defects within starter drive or components

Step 1—Slowly pull on the starter rope, testing that there is a definite resistance being felt to the pull force. When the rope starter is properly engaging the engine crankshaft, a distinct resistance will be felt due to the engine compression. A very light resistance, or no resistance when the rope is pulled slowly can mean:

a.   A defective starter mechanism.

b.   Worn parts on the starter mechanism.

c.   Worn parts on the engine crankshaft.

d.   A loose, slipping starter drive mechanism.

e.   A leaky head gasket.

Was there a definite resistance to the pull on the starter rope?

YES—Proceed to and recheck Fault Symptoms **2321, 2322, 2070.**

NO—Proceed to Step 2.

Step 2—Remove the ignition cable and the spark plug (per Service Procedure **4004**). Pull on the rope pull and listen for a popping sound at the spark-plug hole.

Is there a popping sound which indicates the engine is turning over?

YES—Proceed to Step 3.

NO—Proceed to Step 4.

Step 3—When the rope is pulled, does the rope move out in spurts or jerks, rather than a smooth continuous withdrawal?

YES—Fault has been isolated: Defect internal in starter mechanism. Refer to Service Procedure **4040.**

NO—Proceed to Step 4.

Step 4—Is there any resistance, or very slight resistance felt on the rope pull?

YES—Check compression (Fault Symptom **2500**) and condition of starter mechanism (Service Procedure **4040**) in that order.

NO—If there is no resistance at all the starter mechanism is defective (see Service Procedure **4040**) or a major failure in compression exists (see Fault Symptom **2500**).

ENGINE IS HARD TO START
WIND-UP IMPULSE
STARTER ENGINE

FAULT SYMPTOM   2330

Possible Causes:

- Improper wind-up procedure ——————— **See 2331**

- Wind-up spring is weak ——————————— **See 2332**

- Wind-up spring is defective ———————— **See 2333**

Initial Conditions Check List:

a. Perform the checks given in the Check List of Fault Symptom **1000.**

b. Place the THROTTLE control to OFF or STOP. Turn ignition switch to OFF, if mower is so equipped.

c. Be sure you are following the mower manufacturer's instructions in the use and wind-up of the starter. Be certain you are fully winding the starter so that the starter spring will rotate the engine crankshaft with the full design force. Winding the starter only part of the way may not be sufficient to rotate the crankshaft at the speed needed to generate the required electricity to fire the spark plug.

———— CAUTION————

WHEN WORKING WITH IMPULSE STARTER BE EXTREMELY CAREFUL IN HANDLING THE STARTER ASSEMBLY. IF ASSEMBLY IS REMOVED FROM ENGINE, WEAR SAFETY GLASSES DURING REMOVAL AND SERVICING FOR EYE PROTECTION

———— CAUTION————

2331

Improper wind-up procedure

Step 1—Place the THROTTLE control to RUN or START. Turn ignition to RUN or START. Turn ignition switch to ON, if so equipped. Crank up the wind-up starter handle—to the maximum number of turns it will go (usually about 6 turns). If mower has a separate CHOKE control, place at CHOKE. Release starter.

Did the engine start up easily?

YES—Proceed to Step 2.

NO—Proceed to Step 4.

Step 2—Repeat Step 1 except do not wind starter fully. Wind only about 2 to 3 turns. Release starter.

Did the engine start up easily?

YES—Fault has not been isolated. Proceed to Step 3.

NO—Proceed to Step 4.

Step 3—Repeat Step 1, except wind to 4 to 5 turns. Release starter.

Did the engine start easily?

YES—Fault has been isolated: Improper wind-up.

NO—Proceed to Step 4.

Step 4—Upon release of the wound-up starter, is it evident that the engine is being cranked over? The sound the engine makes is the easiest tell-tale sign.

YES—Proceed to Fault Symptom **2332**.

NO—Proceed to Fault Symptom **2333**.

2332

Wind-up spring is weak

Step 1—The engine may not be cranking fast enough using the wind-up starter. The following check permits this to be determined:

a. See if the impulse starter incorporates a debris screen over the top of the engine magneto flywheel. If so, this screen rotates with the starter, proceed to b. If the screen is inaccessible or there is none, proceed to Step 2.

b. Observe the screen when the starter is wound up and released. A starter which has a good spring, when wound and released will, at the time of release, be able to spin the engine fast enough so that the small holes or openings in the screen are moved too quickly to be seen (when the starter is first released).

Can the debris screen holes be observed when the screen rotates?

YES—Fault has been isolated: Weak wind-up spring. Refer to Service Procedure **4041**.

NO—Engine was cranked fast enough. Therefore, the fault was not isolated. The cause of hard starting cannot be attributed to the wind-up spring. Refer to index for other causes of hard start. Proceed to Step 2.

Step 2—If the debris screen is inaccessible, or there is none on the mower, or the results are inconclusive, then check the engine sound. When the wind-up starter is released, the engine sound will help to evaluate the starter's performance. Test the starter as follows:

a. A small lawn mower engine—when running properly—at even very low speeds makes a putt putt noise.

b. The putt-putt sound can be readily discerned at low speeds, and at cranking speeds.

c. As the engine is cranked faster, or runs faster, these putt putt sounds will occur closer together in time. Eventually they flow together as a continuous sound when the engine runs faster.

d. A simple check of the engine's cranking sound would be to listen to the sound of your neighbor's lawnmower as it is wound up and released, and when it starts up.

Then compare that with your own mower's cranking sound.

Is the sound your mower makes the same as a mower that starts up easily?

YES—The fault is not isolated. Engine was cranked fast enough. Therefore, the fault was not isolated. The cause of hard starting cannot be attributed to the wind-up spring. Refer to index for other causes of hard start.

NO—Proceed to Step 3.

Step 3—Does the lawn mower make a very slow and discernible putt putt sound when cranked?

YES—Fault has been isolated: Weak wind-up spring. Refer to Service Procedure 4041.

NO—If the mower makes no or almost no putt putt sound when the starter is used then proceed to Fault Symptom 2333.

2333

Wind-up spring is defective

Step 1—Wind up the starter as cited under 2331 Step 1.

Does the handle actually wind up the starter?

YES—Proceed to Step 2.

NO—Fault has been isolated: Defective wind-up spring. Refer to Service Procedure 4041.

Step 2—Check the condition of the impulse starter's spring as follows:

a. Wind the starter only part way—about 1½ turns to 2 turns (most wind-up starters require about 6 or more turns to be fully wound).

b. Release starter.

The engine should crank over, even with this partial wind up.

Did the engine crank at all?

YES—Proceed to Fault Symptom 2500.

NO—The fault has been isolated: Wind-up spring is disconnected or broken. Refer to Service Procedure 4041.

ENGINE IS HARD TO START CHECK FOR LOOSE ROTARY BLADE

FAULT SYMPTOM   2400

Possible Cause:

- Rotary engine blade is loose

Initial Conditions Check List:

a. Perform the checks given in Check List of Fault Symptom **1000.**

b. This test applies only to lawn mowers which have the rotary blade mounted directly on the engine crankshaft. For mowers which do not have the blade mounted directly on the crankshaft, proceed to Fault Symptom **2500** and check index for other causes of hard start.

```
┌──────────── CAUTION ────────────┐
│                                  │
│     BEFORE PROCEEDING            │
│     DISCONNECT IGNITION          │
│     CABLE FROM SPARK             │
│     PLUG, REMOVE PLUG,           │
│     (PER SERVICE PROCED-         │
│     URE 4004), SET THROT-        │
│     TLE CONTROL TO OFF—          │
│     USE CARE IN HANDLING         │
│     THE MOWER CUTTING            │
│     BLADE                        │
│                                  │
└──────────── CAUTION ────────────┘
```

Step 1—Tilt the mower up so that the cutting blade can be examined. Carefully check if the blade is securely attached to the engine crankshaft:

a. Attempt to move the blade up and down. If the blade is properly secured, it will not be possible to move the blade.

b. A secure blade will not move at all, relative to the engine crankshaft.

c. Attempt to rotate the blade slowly, and see if there is any movement between the blade hub and the engine crankshaft.

Is the blade secure?

YES—Proceed to Fault Symptom **2500.**

NO—Refer to Service Procedure **4002,** for blade tightening, and then proceed to Step 2.

Step 2—Return mower to operating position, replace spark plug, replace ignition cable and start up.

Did the engine start up easily?

YES—Fault has been isolated and verified: Loose rotary blade.

NO—Proceed to Fault Symptom **2500.**

ENGINE IS HARD TO START
CHECK ENGINE COMPRESSION
MECHANICAL CONDITION

FAULT SYMPTOM   2500

Possible Causes:

- No oil in crankcase (4-cycle engine) ——————— **See 2510**
- Loose cylinder head bolts ——————————— **See 2520**
- Warped cylinder head ————————————— **See 2530**
- Blown/defective cylinder head gasket ————— **See 2540**
- Worn/defective piston rings or cylinder walls —— **See 2550**

Initial Conditions Check List:

a. Perform the checks given in Check List of Fault Symptom **1000** and Check List of Fault Symptom **2000**.

After performing the checks indicated, attempt engine start up.

```
———————— CAUTION ————————

        KEEP HANDS, FEET
        CLEAR OF MOWER
        BLADE(S) AND DECK
        BEFORE STARTING

———————— CAUTION ————————
```

Does the engine start up easily?

YES—Fault not verified. Probable oversight in start up procedures.

NO—Proceed to Fault Symptom **2510**.

2510

No oil in crankcase (4 –cycle engine)

Step 1—Check the oil level in the engine crankcase.

Is there the proper amount of oil in the crankcase?

YES—Proceed to Fault Symptom **2520**.

NO—Proceed to Step 2.

Step 2—Fill the crankcase with engine oil per the manufacturer's instructions. Start up the engine.

Did the engine start up easily?

YES—Fault has been isolated: Insufficient oil in engine. Lack of oil gave low compression and hard starting. Check for cause.

NO—Proceed to Fault Symptom **2520**.

2520

Loose cylinder head bolts

Step 1—Remove the ignition cable from the spark plug and place THROTTLE control at STOP or OFF. Check the cylinder head bolts as follows:

a. Slowly crank the engine.

b. Listen carefully for a hissing sound near the cylinder head of the engine (near that end of the engine where the spark plug is located).

c. If a hissing sound is heard, check the bolts which hold the engine cylinder head to the engine block (see Fig. **51**).

d. A bolt tightness test is very simple. Use a wrench which exactly fits the bolt head (preferably a box wrench or a socket wrench).

e. Hold the wrench (or the socket wrench handle) somewhere in the middle of its length. See Fig. **51**. (Do not hold the wrench at its far end—opposite from where the bolt is—since this may provide too much leverage.)

f. Slowly turn the wrench clockwise. Use a light force; do not exert a heavy force! If the bolt is properly secured, it should not turn.

g. Check all of the bolts on the head in this manner.

Were any of the bolts loose?

YES—Proceed to Step 2.

NO—Proceed to Step 3.

Step 2—A loose head bolt (or several loose bolts) will require tightening for proper engine operation. Refer to Service Procedure **4060** for proper tightening of these head bolts and tightening to the required foot pounds of torque.

Upon completion of Service Procedure **4060** bolt tightening, attempt start up.

Did the engine start easily?

FIGURE 51   Cylinder Head Bolt Tightness Check

YES—Fault has been isolated: Loose cylinder head bolts.

NO—Proceed to Fault Symptom **2530**.

Step 3—Was a hissing sound heard near the cylinder head as the engine was cranked?

YES—Fault has been isolated: Warped head or blown gasket. Proceed to Fault Symptom **2530** and **2540**.

NO—Proceed to Fault Symptom **2550**.

2530

Warped cylinder head

Step 1—When the cylinder head bolts are tight and a hissing sound is heard, the cylinder head should be checked for warpage as follows:

a. Clean the area around the engine head.

b. Carefully examine the joint or area where the head is joined to the rest of the engine cylinder.

c. This joint should form a straight even line all around the cylinder head. At no point should it be wavy or show bumps.

Is the joint between the cylinder head and the engine uniform?

YES—Proceed to Fault Symptom **2540.**

NO—Fault has been isolated: Warped cylinder head. Refer to Service Procedure **4060** for removal and servicing.

2540

Blown/defective cylinder head gasket

Step 1—Re-examine the cylinder head as follows:

a. Look for any evidence of gaps in the gasket joint which forms the even line between the cylinder head and the engine block.

b. The width of the joint should be the same all around.

c. Gray/black streaks along the gasket line indicate the location of the leak.

Does the cylinder head have any of these signs?

YES—Fault has been isolated: Blown head gasket. Refer to Service Procedure **4060** for gasket replacement.

NO—Proceed to Fault Symptom **2550.**

2550

Worn/defective piston rings or cylinder walls

Step 1—The tests in **2510** to **2540** are gross tests to quickly determine obvious compression problems. Engine compression should be measured with a compression gage to conclusively establish the existence of a compression problem. Typical gages are shown in Figs. **52** and **53**. To perform this test, remove the spark plug and insert the gage in the plug hole. Be sure the gage fits tightly in the spark-plug hole, adjust gage as shown in Fig. **54.** Proceed as follows:

a. Place the THROTTLE control to FAST (or RUN or HIGH).

FIGURE 52   Compression Gage Tester, with Handle

FIGURE 53   Compression Gage Tester, Plug-in Type

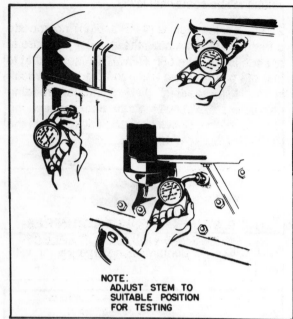

NOTE:
ADJUST STEM TO
SUITABLE POSITION
FOR TESTING

FIGURE 54   Use of Compression Test Gage

b.   Use the starter to crank the engine, crank as fast as possible.

c.   Read the dial on the gage when the needle is at the highest point.
•   Make a number of readings.
•   Average the readings taken by dividing the sum of the values by the number of readings taken (example: four readings, 90+110+90+110=400; 400 ÷ by 4 = 100).

d.   Proceed to Step 2.

Step 2—Refer to Table II for normal compression values for typical engines. If your engine is not listed, obtain the data from an authorized dealer.

Compare the readings and observe the behavior of the gage to evaluate the condition of the engines as follows:

a.   Is the average value of the readings ½ or less than ½ the value given in Table II for good compression?

YES—Fault has been isolated: Bad compression. Refer to Service Procedure **4100.**

NO—Proceed to "b" for 4-cycle and "e" for 2-cycle engines.

b.   Observe the action of the gage pointer. If the gage pointer does not climb steadily but goes to a low reading of perhaps 25 psi and stays there, then for 4-cycle engines this indicates the exhaust valve is not closing fully.

Does the gage go to a low reading and not climb steadily to a high value?

YES—Fault has been isolated: Open exhaust valve. Refer to Service Procedure **4100.**

NO—Proceed to "c."

c.   Observe the gage pointer. If the pointer goes to 25 psi, then climbs to 30 or 40 and then goes down to 25, this indicates that there is a valve sticking. A sticky valve will probably show up on the first few revolutions of the engine.

Does the pointer oscillate up and down as noted?

YES—Fault has been isolated: Sticky valve. Refer to Service Procedure **4100.**

NO—Proceed to "d."

d.   If the gage shows a very low (10 psi) reading or no reading, then it is probable that a valve is stuck fully open (for 4-cycle engines) or the piston head may have a hole burned through it (for both 2- and 4-cycle engines).

Does the gage show a very low reading?

YES—Fault has been isolated: Stuck open valve or defective piston. Refer to Service Procedure **4100.** As a final check on worn piston rings or internal wear in the cylinder, proceed to "g."

NO—Proceed to "e."

e.   For 2-cycle engines if the gage reading is low or less than half the value given, then it is most likely that the piston rings are worn or there is internal wear or scouring on the cylinder walls.

Is the gage reading less than ½ the value given for good compression?

YES—Fault has been isolated: Piston rings worn or damaged cylinder walls. Refer to Service Procedure **4100.**

NO—Proceed to "f."

f. Special note should be taken if the compression pressure reading is very much higher than the table value (by 15 or more psi). This indicates that carbon deposits have built up inside the cylinder and trouble can be expected in the near future.

Is the gage value much higher than Table II value?

YES—Refer to Service Procedure **4060.**

NO—If the readings taken are about the same as those given in Table II, proceed to Fault - Symptom **2600** for 2-cycle engines, or **2000** for 4-cycle engine hard-to-start symptoms.

g. Before reassembly of the spark plug into the engine a simple final check for worn rings or cylinder walls can be made. Pour about a teaspoon of engine oil into the cylinder through the spark plug hole. Crank the engine once or twice to distribute the oil, then take another compression reading. The oil will temporarily seal off the worn rings or scuffed/damaged cylinder walls and give a higher compression reading. This is only temporary; when the oil drains down into the cylinder the readings will be low again.

Did the gage show a higher compression immediately after adding oil, then a low reading later?

YES—Fault verified. Piston rings worn or damaged cylinder walls. Refer to Service Procedure **4100.**

NO—If the gage still showed low readings the rings or walls are badly worn. Refer to Service Procedure **4100.** If the gage showed high/normal readings which stayed high, refer again to Fault Symptom **2600** for 2-cycle engines, or **2000** for 4-cycle engines.

NOTE:

1. These values are for a warm engine with the THROTTLE wide open. If the engine is cold, the values will be somewhat lower.

2. The Kohler engines (4-HP and up) incorporate a feature which releases the cylinder pressure at speeds lower than 650 RPM. This means that to test the compression using the starter will result in a false reading. For a proper reading disconnect the mower blade and connect the shaft to the chuck of an electric drill to obtain the desired higher speed.

TABLE II

TABULATION OF CYLINDER COMPRESSION PRESSURE VALUES FOR SELECTED LAWN MOWER ENGINES

Engine Manufacturer and Model Number	Compression Pressure reading in Pounds per Square Inch (psi)
Jacobsen Model 321 engine	75 - 85
Kohler (see note 2)	
K 91 K181 K141 K241 K161 K301	110 - 120
Clinton 2 Stroke (or 2 Cycle)	60 or higher
up to 4.5 HP	65 - 70
over 4.5 HP	70
Tecumseh Model AH 750	100
AC 750	100
Model AH 520	
AV 520	
AV 600	
AH 440	90
AH 480	
AH 490	

ENGINE IS HARD TO START
2-CYCLE ENGINES—
MECHANICAL CONDITION
FAULT SYMPTOM 2600

Possible Causes:

- Leaky crankcase seal—Rotary Mower Engines —— **See 2610**

- Leaky crankcase seal—Reel Mower Engines —— **See 2620**

- Leaky ignition seal—Reel or Rotary

  Mower Engines——————————————— **See 2630**

- Blocked exhaust ports ———————————— **See 2640**

- Defective Reed valve ———————————— **See 2650**

Initial Conditions Check List:

a. Perform the checks given in Check List of Fault Symptom **1000** and Check List of Fault Symptom **2000**.

b. The procedures given for check of a leaky crankcase seal on rotary mower engines apply to engines which have the cutting blade mounted directly on the crankshaft. For rotary mower engines which do not have the blade so mounted, follow the procedures given for reel mower engines.

```
———— CAUTION ————


      KEEP HANDS, FEET
      CLEAR OF MOWER
      BLADE(S) BEFORE
         STARTING


———— CAUTION ————
```

2610

Leaky crankcase seal—rotary mower engines

Step 1—Proceed as follows to test for a leaky crankcase seal on mowers which have the cutting blade mounted on the engine crankshaft.

a. Place the THROTTLE control to OFF or STOP, and disconnect the spark-plug ignition cable.

b. Tilt the mower up, being careful not to have the fuel tank or oil port spill over.

c. Examine the cutting blade and the cutting blade adapter (see Fig. 55).

Look for evidence of raw unburned fuel or oil on these parts. In normal operation these parts should have a dry gray to black exhaust gas film covering them.

If the crankcase seal is leaking, raw fuel and oil should be found on the blade and/or adapter.

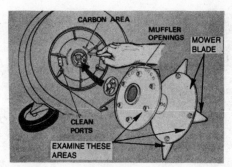

FIGURE 55 Examination of Cutting Blade Areas for Raw Unburned Fuel

FIGURE 56 Leaky Crankcase Seal Test

In the case of a 2-cycle rotary lawn mower engine which has the exhaust port discharge under the deck a leaky crankcase seal is most likely to deposit raw fuel-oil mixture on the blade and adaptor.

Is there evidence of raw fuel or oil on the blade and/or adaptor?

YES—Fault has been isolated: Leaky crankcase seal. Refer to Service Procedure **4100.**

NO—Proceed to Fault Symptom **2630.**

2620

Leaky crankcase seal—reel mower engines

Step 1—A leaky crankcase seal on a reel mower engine can be checked very simply and quickly. Most reel mowers mount the engine so that the crankshaft is horizontal. Therefore the crankshaft seal for these engines is readily accessible.

If the engine has been running, wait until it has completely cooled before performing this test.

Reel mowers, or mowers which have the engine crankshaft mounted horizontally have the crankcase seal on the side where the engine power take off is located. See Fig. **56.** The power take off usually drives a gear reduction box or a belt pulley, or a sprocket/chain pulley.

Proceed as follows:

a. The engine must be started up.

b. Run the engine in as low an RPM or speed as possible.

CAUTION

BE ABSOLUTELY CER-
TAIN NOT TO ENGAGE
THE CLUTCH OR POWER
TAKE OFF—BE EX-
TREMELY CAREFUL
WITH HANDS OR
LOOSE GARMENTS—
SO AS NOT TO GET
CAUGHT IN ANY
MECHANISM—IF IT
IS NECESSARY TO
REMOVE ANY BELT
OR CLUTCH GUARD
TO GET AT THE
CRANKCASE SEAL,
USE A LONG SPOUT
OIL CAN, OR A LONG
WOODEN ARTIST'S
BRUSH TO PUT OIL
AT SEAL

CAUTION

---

c. Refer to Fig. 56. Squirt or place a small amount of engine oil around the crankcase seal.

d. Listen carefully to the sound of the engine.

Has the sound of the engine changed to a smoother, higher pitch?

YES—Fault has been isolated: Leaky crankcase seal. Refer to Service Procedure **4070.**

NO—Proceed to Step 2.

Step 2—Look at the exhaust muffler as you squirt oil on the seal.

Is there a marked increase in the amount of gray/white smoke?

YES—Fault has been isolated: Leaky crankcase seal. Refer to Service Procedure **40 70**.

NO—Proceed to Fault Symptom **2630**.

2630

Leaky ignition seal—reel or rotary mower engines

Step 1—The engine has an oil seal which is at the top end of the crankshaft, at the flywheel end. This seal is usually called the ignition seal since it seals off the crankshaft from the ignition system which is mounted all around the seal. The seal (and the ignition system) is located under the magneto flywheel (see Fig. 57). It is necessary to first remove the flywheel in order to inspect or service this seal.

Refer to Service Procedure **4035** for the steps necessary to remove the flywheel and to check the ignition seal.

Upon removal of the magneto flywheel, proceed as follows:

a. Examine the ignition parts around the seal. Look for signs of oil or raw fuel on the parts.

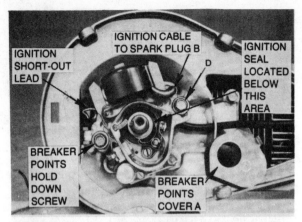

FIGURE 57 Engine Flywheel Magneto Removed Showing Ignition System Components and Location of Ignition Seal

b. Run your finger lightly over the parts and at the base of the crankshaft.

Was there any sign of oil or raw fuel around the ignition seal?

YES—Fault has been isolated: Leaky ignition seal. Refer to Service Procedure **4035**.

NO—Service the ignition system per Service Procedure **4030** and refer to Fault Symptom **1000** or **2000** for other system checks for hard-starting.

ENGINE IS HARD TO START 2-CYCLE ENGINES—BLOCKED EXHAUST PORTS

FAULT SYMPTOM   2640

Initial Conditions Check List:

Perform the checks given in Fault Symptom **1000** and Fault Symptom **2000**.

2641

Horizontally mounted 2-cycle engines—horizontal crankshaft

Step 1—In a 2-cycle engine the intake of fuel-air mixture and the exhaust of the burned gases are accomplished by the use of a special valve and intake and exhaust ports. These ports are built into the walls of the cylinder as shown in the sketch of a cross-section of an engine, Fig. 58. Engines which are mounted horizontally, with the crankshaft horizontal, incorporate a muffler which is part of the cover over the exhaust ports.

If carbon is permitted to build up at the exhaust ports to the point where the ports are becoming blocked, then the engine will be hard to start. See Fig. 59. Proceed as follows to check the exhaust ports:

```
――――――― CAUTION ―――――――

        IF ENGINE IS HOT
      WAIT UNTIL ENGINE
       COOLS BEFORE
          PROCEEDING

――――――― CAUTION ―――――――
```

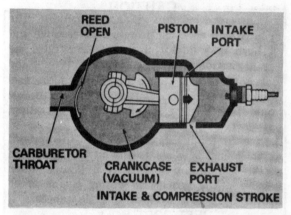

FIGURE 58   Cross-section of Typical 2-Cycle Engine

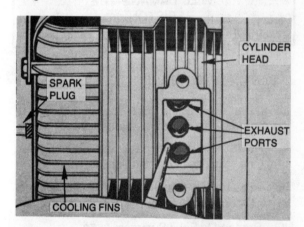

FIGURE 59   Two-Cycle Engine Exhaust Ports on Horizontal Engine, Muffler Removed

a. Place controls or THROTTLE at OFF or STOP. Remove ignition cable from spark plug.

b. Remove the cover plate/or muffler over the engine exhaust ports. On horizontally mounted engines the exhaust ports/muffler will usually be on the side of the cylinder. Use care to select the correct wrench; place the cover and bolts in a container for safe keeping.

c. If carbon is present, use a narrow paint scraper, or knife with a squared blunt end blade to scrape the carbon from the exhaust port openings.

CAUTION

BE CERTAIN THAT THE CYLINDER PISTON IS NOT RAISED UP TO THE LEVEL OF THE EXHAUST PORT HOLES WHEN CLEANING AROUND THE HOLES— USE THE STARTER TO LOWER THE PISTON TO BELOW THE LOW- EST PORT HOLE— SCRAPING THE PISTON WILL DAMAGE THE PISTON

CAUTION

---

d. Replace the exhaust port cover and/or muffler. Be certain bolts are tight.

e. Reconnect ignition cable and attempt engine start up. Did the engine start easily?

YES—Fault has been isolated and verified: Clogged exhaust ports.

NO—Proceed to Fault Symptom **2650.**

2642

Vertically mounted 2-cycle engines—vertical crankshaft

Step 1—On 2-cycle engines which are mounted vertically, with the crankshaft vertical, the exhaust ports are usually on the bottom or underside of the engine cylinder. This permits the exhaust gases to be deflected below the deck of the mower and onto a circular muffler. See Fig. 60. Check the exhaust ports and clean as required.

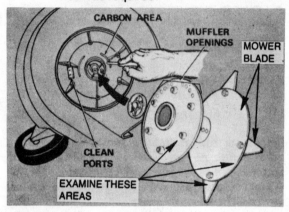

FIGURE 60   Sketch of Circular Muffler and Bottom of Mower Deck

---

CAUTION

IF ENGINE IS HOT WAIT UNTIL IT COOLS BEFORE PROCEEDING

CAUTION

---

a. Place controls on THROTTLE at STOP or OFF. Remove ignition cable from spark plug.

b. Tilt the mower deck up so as to gain access to the bottom of the deck. Do not tilt to the extent that oil or fuel will be spilled.

c. Remove the lawn mower blade (refer to Service Procedure **4002**).

d. Remove the bolts which are holding the muffler and place in a container for safe-keeping.

e. Remove the exhaust port cover, if there is one.

f. Check for carbon in the exhaust ports. If present, use a narrow paint scraper or knife with a squared blunt blade to remove it.

─── **CAUTION** ───

BE CERTAIN THAT THE CYLINDER
PISTON IS NOT RAISED UP TO
THE LEVEL OF THE EXHAUST PORT
HOLES WHEN CLEANING AROUND THE
HOLES— USE THE STARTER TO
LOWER THE PISTON TO BELOW THE
LOWEST PORT HOLE—SCRAPING THE
PISTON WILL DAMAGE THE PISTON

─── **CAUTION** ───

g. After cleaning out the carbon from the exhaust ports, also clean out any carbon deposits from the muffler openings and from the mower deck. Carbon will tend to accumulate on and around any ridge or depression in the deck. The deck area should be scraped free of all carbon.

h. Replace the exhaust port cover and/or muffler. Be certain bolts are tight and be certain to include any lock washers which were under the bolt heads. Replace the blade and be sure the blade bolts are tight. Restore the mower to a normal position.

i. Reconnect the ignition cable to spark plug and start up the engine.

─── **CAUTION** ───

KEEP HANDS, FEET
CLEAR OF MOWER
BLADE(S) BEFORE
STARTING

─── **CAUTION** ───

Did the engine start easily?

YES—Fault has been repaired: Hard-starts caused by clogged exhaust ports.

NO—Proceed to Fault Symptom **2650**.

2650

2-cycle engines—defective reed valve

Step 1—A major moving part of a 2-cycle engine is the reed valve. This valve controls the flow of the fuel air mixture into the crankcase of the engine as shown in the sketch of Fig. 61. The reed valve opens and closes in accordance with the movement of the engine piston as shown in Figs. 62, 63 and 64. It is evident from these sketches that if the valve does not work properly the engine will not perform well. In addition, if the valve is sticky, leaking, or broken, the engine will be hard to start. A typical reed valve which has been removed from the engine is shown in Fig. 65. Test the reed valve as follows:

a. Place controls or THROTTLE at OFF or STOP.

b. Remove the carburetor air cleaner assembly. See Fig. 66.

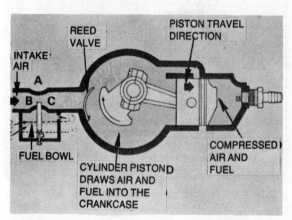

FIGURE 61   Typical Reed Valve Operation, 2-Cycle Engine

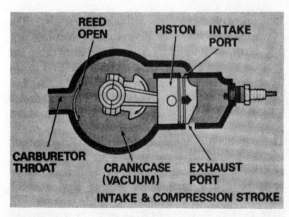

FIGURE 62   Reed Valve Opens, Admits Fuel/Air

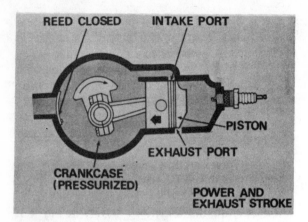

FIGURE 63   Reed Valve Closes

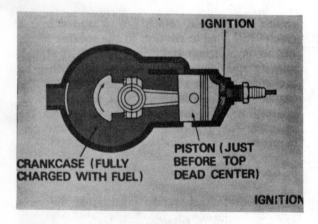

FIGURE 64   Ignition Phase

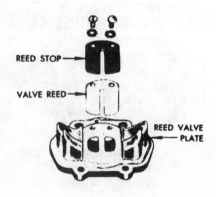

FIGURE 65   Typical Reed Valve and Adapter
Plate

FIGURE 66   Test for Leaky Reed Valve

```
━━━━ CAUTION ━━━━
    KEEP HANDS, FEET
CLEAR OF MOWER BLADE(S)
    BEFORE STARTING
━━━━ CAUTION ━━━━
```

c.   Start the engine (follow Start-up Procedure **1000**).

d.   With the engine running, hold a 1½ to 2 inch wide piece of clean white paper about 1 inch from the intake of the carburetor as shown in Fig. 66. If the engine cannot be started, proceed to Step 2.

e.   Examine the paper.

Has the paper been spotted by the fuel mixture?

YES—Fault has been isolated: Leaky or bad reed valve. Refer to Service Procedure **4012**.

NO—Refer back to Fault Symptom **2000** for other tests which may be performed for hard starting.

Step 2—If the engine cannot be started, perform the test by having someone rapidly activate the starter 4 to 5 times. Hold the paper closer, about ½ inch away in this case.

Has the paper been spotted by the fuel mixture?

YES—Fault has been isolated: Leaky or bad reed valve. Refer to Service Procedure **4012**.

NO—Refer back to Fault Symptom **2000** for other tests which may be performed for hard starting.

ENGINE RUNNING PROBLEMS FAULT SYMPTOM 3000

Symptoms:

- Engine sputters, recovers, sputters, slows and stops ————————————— **See 3010**
- Engine slows, shows difficulty in running, finally stops ————————————— **See 3020**
- Mower has difficulty in cutting high grass ———— **See 3030**
- Engine lacks speed control capability ————— **See 3040**
- Engine sound, abnormally loud or quiet ———— **See 3050**
- Engine gives off excessive smoke ————— **See 3060**
- Engine vibrates excessively ————————— **See 3070**
- Engine misfires ————————————— **See 3080**
- Engine backfires ————————————— **See 3090**

Initial Conditions Check List:

a. Perform all the pre-start checks given in Fault Symptom **1000** and Fault Symptom **2000**.

3010

Engine sputters, recovers, sputters, slows and stops

Step 1—Start and run the lawnmower engine. Place the THROTTLE control at RUN, or midposition. Leave the control at this position for about 3 to 5 minutes, and then proceed as follows:

---CAUTION---

USE GREAT CARE
WHEN OPERATING
THE MOWER DURING
THESE TESTS.
KEEP HANDS AND
FEET CLEAR OF
MOWER BLADE(S)
AND ANY MOVING
OR ROTATING PART.
DO NOT PERFORM ANY
TEST ON A HILLSIDE
OR A GRASSY SLOPE.
SELECT A FLAT AND
LEVEL LAWN OR
GRASS AREA FOR
MAXIMUM SAFETY
OF OPERATION.

---CAUTION---

a. While the engine is running, listen to the sound it makes. A lawn mower engine should be able to run smoothly and consistently at this speed without need of continuous movement of the THROTTLE control. The engine should give an even, constant, buzzing sound.

b. If the engine runs for a while, then sputters, recovers, runs for a while, again sputters, and finally slows down to a gradual stop, there are problems in the fuel system.

Did the engine sputter, run, sputter and then slow down to a gradual stop?

YES—A possible cause is a faulty fuel system. Proceed to Fault Symptom **3100.**

NO—Proceed to Step 2.

Step 2—Leave the THROTTLE at the same position as in Step 1. Permit the engine to run for another 5 to 10 minutes.

During this time (or at any time when the engine was otherwise considered to be working properly), did the engine come to a sudden, abrupt stop?

YES—Fault due to use of wrong fuel, engine overheating, or major mechanical fault. Proceed to Step 3.

NO—Proceed to Fault Symptom **3020.**

Step 3—If the engine did stop abruptly, permit it to cool down. Restart the engine.

Did the engine restart?

YES—Proceed to Fault Symptom **3200** for verification of wrong fuel or engine overheating.

NO—Engine may have a major mechanical internal fault. Refer back to Fault Symptom **1000** to determine if engine will crank.

3020

Engine shows difficulty in running, finally stops

Step 1—Repeat the test of **3010,** Step 1 a and b.

While the engine was running did the engine run roughly, with more than usual vibration—almost bouncing along—and not give a smooth buzzing sound?

YES—Proceed to Step 2.

NO—Proceed to Fault Symptom **3030.**

Step 2—Did the engine slow down, exhibit difficulty in running, or finally slow and stop?

YES—A possible cause: ignition is faulty. Proceed to Fault Symptom **3200.**

NO—Proceed to Fault Symptom **3030.**

3030

Mower has difficulty in cutting high grass

Step 1—Shut off the lawn mower engine if it is running. Proceed with mower cutting power test as follows:

a. Set the grass cut height to give about 1½ inches height of grass, or at the halfway position of the height control (usually located at the wheels for rotary mowers, and usually at the roller bar for reel mowers).

b. Select an area of high dense grass for this test (grass about 10 inches or so high).

CAUTION

IF THE MOWER BLADE IS DULL, WORN, OR BADLY IN NEED OF SHARPENING, THEN THE MOWER ENGINE MAY BE STALLED AND STOPPED WHEN TRYING TO CUT THIS VERY HIGH GRASS. PERFORM THIS TEST WITH A SHARP BLADE IN GOOD CONDITION. REFER TO PROCEDURE **4002** FOR BLADE SHARPENING.

CAUTION

c. Start and run the mower engine.

d. Place the THROTTLE control at HIGH, or FAST.

e. Cut this high grass.

f. If the engine is operating properly and has good power, the lawn mower should be able to cut this high grass. The mower engine should not be slowed down so much that it stops, although the high speed may not necessarily remain high while the high grass is being cut. The mower speed will depend on how fast the mower is moved through the high grass.

g. This test is a measure of the power of the engine, and its capability to sustain high power.

Did the lawn mower successfully cut the high grass?

YES—Proceed to Step 2.

NO—The fault has been verified: Mower engine lacks power. Proceed to Fault Symptom **3400.**

Step 2—Allow the mower to run, positioning the THROTTLE at midposition or RUN. Then cut an area of grass which is a normal 3 or 4 inches in height.

Did the mower cut this grass easily?

YES—Proceed to Fault Symptom **3040.**

NO—Fault has been verified: Mower engine lacks power. Proceed to Fault Symptom **3400.**

3040

Engine lacks speed control capability

Step 1—To operate properly, it is necessary to have control of the engine at all speeds, high, medium, and low or idle. Perform the following test:

a. If the engine is not running, start up, and run the engine.

b. Disengage any power drive equipment as the reel, self-propelling mechanism, clutches or sprockets.

c. Position the THROTTLE control at the IDLE position.

d. Move the control back and forth. The engine should speed up and slow down accordingly.

e. Maintain the control at IDLE.

Did the engine IDLE?

YES—Proceed to Step 2.

NO—Fault has been isolated: Engine IDLE not possible. Refer to Fault Symptom **3500.**

Step 2—When the control is positioned and left at a given setting, the engine speed should stay at that speed setting. Perform the following test:

a. Operate the mower speed control.

b. It may be noticed that when the THROTTLE control is placed at a given setting, or about at midposition, the engine does not maintain that speed. The engine may speed up, slow down and then repeat speeding up and slowing down.

c. When the engine does this without the control being moved, this condition is called engine surge.

d. Engine surge is very undesirable since speed control cannot be maintained.

e. This condition can be caused by faults:
- in the fuel system.
- the engine speed governor system.
- in other systems.

Did the engine give any indication of engine surge?

YES—Fault has been isolated: Engine speed surge occuring. Refer to Fault Symptom **3600.**

NO—Proceed to Fault Symptom **3050.**

3050

Engine sound, abnormally loud or quiet

Step 1—With the mower engine running, move the THROTTLE control back and forth from low to high speed and back to low. As you move the control perform the following:

a. Listen to the engine sound. When the engine speeds up, the sound should get louder, then lower when the speed is reduced.

b. The engine sound at high speed should be a tolerable sound. The sound the engine makes should not be a very high roar which can be heard a long distance away.

c. Additionally, the sound the engine makes should not suddenly become very loud—to a roaring, sharp banging sound.

d. If this noise is heard, then the exhaust system or the muffler has a defect. The muffler may be blown out internally, or it may be cracked. Two-stroke (2-cycle) engines require special attention as discussed in Step 2 which follows.

Did the lawn mower engine make any of the loud sounds described?

YES—Fault has been isolated: Defect in exhaust system. Proceed to Fault Symptom **3700**.

NO—Proceed to Step 2.

Step 2—The 2-cycle engine may become progressively quieter in operation, up to that point where a soft, quiet, exhaust sound is heard. If this occurs, it means the muffler is plugged up with carbon deposits, or the engine exhaust ports are clogged up with deposits.

Has the mower engine become progressively quieter in operation?

YES—Fault has been isolated: Muffler becoming plugged or exhaust ports plugged. Proceed to Fault Symptom **3700**.

NO—Fault not verified. Proceed to **3060** for a related test.

3060

Engine gives off excess smoke

Step 1—Some very simple checks can be made on the engine exhaust condition. When an engine is operating properly, there should be very little or no exhaust gas color of any kind. The exhaust gas should be colorless when the engine is operating in the midspeed range. Perform the following tests with the mower engine running:

a. Position control at both IDLE, and at wide open THROTTLE. The exhaust should be barely noticeable.

b. When the engine has a faulty carburetor, or is using the wrong fuel, or has a leaky oil seal, the exhaust gas will be very noticeable and smoky.

c. The smoke can be dark gray to black in color or even gray white in color.

d. A poorly running engine can give off exhaust smoke in the form of heavy, thick clouds.

e. If the fuel mixture is improperly adjusted, there can be a strong gasoline smell to the exhaust smoke. If the engine is burning oil, the exhaust smoke will have a strong smell of burned oil. If this is the case, the exhaust will have a sharp caustic smell.

Are any of these signs or odors evident when the mower engine is running?

YES—There may be multiple faults. Refer to Fault Symptoms **3700**.

NO—Fault not verified. Proceed to **3070**.

Step 2—Permit the engine to run and warm up. Then run the following test:

a. Operate the lawn mower THROTTLE control and run the engine at IDLE for a few minutes.

b. Then at a midrange speed position for a few minutes.

c. Then for a few minutes at high speed.

d. Then as fast as the engine will run.

During this test, did the engine exhaust give off excessive smoke of a nature as noted in Step 1 d of **3050**?

YES—Fault has been isolated: Engine in need of internal mechanical repair. Refer to Service Procedure **4100**.

NO—Fault not verified. Proceed to Fault Symptom **3070**.

3070

Engine vibrates excessively

Step 1—When the lawn mower has been in service for one or more seasons, the mower engine may begin to vibrate and even vibrate excessively. The engine may literally move around in relation to the deck or platform to which it is mounted. This fault, if ignored, can lead to a very dangerous condition for the mower operator, since the engine turns the blade at high speeds.

Does the engine vibrate—move around in relation to the mower deck—when the mower is in operation?

YES—Excessive vibration fault. Refer to Fault Symptom **3930**.

NO—Fault not verified. Proceed to Fault Symptom **3080**.

3080

Engine misfires

Step 1—An occasional problem with lawn mowers is having engines misfire, or skip firing while running. The engine should run smoothly and consistently, without any sign of hesitation. When the engine is missing, or misses, this means that the fuel charge in the engine cylinder is not being ignited with every cycle the engine makes. As a result the power which the engine develops will vary

each time the engine misses, since in that cycle there was no power developed. This symptom can be the forerunner of a much more serious problem developing in the engine since misfire can be due to a faulty fuel system and/or faulty ignition system. Proceed to test the engine as follows:

a. With the engine running, operate the lawnmower THROTTLE.

b. Run the engine at IDLE, up to high speed, then back to IDLE by moving the THROTTLE control.

c. Engage, then disengage any power attachments as the reel, clutches, self-propelling mechanisms, pulley or sprocket.

d. Move the THROTTLE back and forth when the power attachment is engaged.

e. In all of these actions the engine should respond to the THROTTLE position, running faster or slower as the THROTTLE is positioned accordingly.

f. There should be a smooth and complete response to this speed control with no evidence of engine hesitation, shudder, or sluggish action.

Did the engine give any indication of misfiring?

YES—Fault has been isolated: Engine misfire. Proceed to Fault Symptom **3800**.

NO—Fault not verified. Proceed to Fault Symptom **3090**.

3090

Engine backfires

Step 1—A companion problem to misfire is engine backfiring. Engine backfiring can be alarming and is also dangerous since the fuel charge is igniting outside of the engine cylinder. If the fuel charge ignites inside the muffler it can blow off the muffler and possibly cause the mower operator injury. Backfiring is a loud, claplike noise easily heard when it occurs. This sound is completely different from the usual steady hum or putt-putt sound the engine makes. If backfiring occurs, discontinue mowing, shut off the engine and begin troubleshooting the engine. The problem can be caused by a faulty fuel mixture or can be more significant—a fault involving major mechanical problems within the engine. Under no circumstances should backfiring be ignored. The engine can be totally ruined, or become dangerous to operate.

Did the engine backfire?

YES—Fault has been verified: Proceed to Fault Symptom **3900**.

NO—Fault not verified. All of this section's running problems have been reviewed in Fault Symptoms **3010** through **3090**. Refer to specific procedures as cited for more detailed information, testing, or correction.

ENGINE RUNNING PROBLEMS
FAULTY FUEL SYSTEM

FAULT SYMPTOM   3100

Possible Causes:

- Water, dirt or rust contamination in fuel ———— **See 3110**

- Fuel vent cap plugged ————————————— **See 3120**

- No fuel supply ——————————————— **See 3130**

- Carburetor adjustment ———————————— **See 3140**

- Fuel starvation from crankcase air leaks ———— **See 3150**

- 2 cycle engines—fuel intake valve defects ———— **See 3160**

Initial Conditions Check List:

a. Perform all the pre-start checks given in Fault Symptom **1000** and Fault Symptom **2000.**

b. If engine is hot, wait until it cools.

3110

Water, dirt or rust contamination in fuel

Step 1—Determine if water, dirt or rust has accummulated in the fuel tank. Refer to Figs. 67, 68, 69, 70, 71 for usual locations of fuel tank drain valves. Proceed as follows:

a. Open the drain valve and drain some fuel from the tank into a wide mouth clean transparent glass jar as shown in Fig. 72.

```
———— CAUTION ————

DO NOT ATTEMPT TO
DRAIN OR FILL FUEL
TANK IF ENGINE IS
HOT. WAIT UNTIL
ENGINE HAS COOLED
TO PREVENT POSSIBLE
FIRE IF FUEL SPILLS
ONTO ENGINE.

————CAUTION————
```

b. Fill the jar about 1/3 full.

c. Shut off the drain valve.

d. Examine the fuel in the jar by holding the jar at eye level.

FIGURE 67 Typical Fuel Bowl Drain Valve
Location

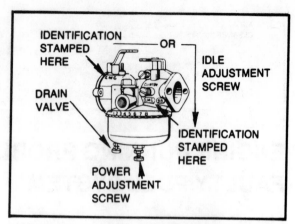

FIGURE 69 Type LMG Carburetor

A. Throttle Plate and Throttle Shaft Assembly
B. Idle Speed Regulating Screw
C. Idle Fuel Regulating Screw
D. Choke Plate and Choke Shaft Assembly
E. Atmospheric Vent Hole
F. Float Bowl Housing Drain Valve
G. Float Bowl Housing Retainer Screw
H. High-speed Adjusting Needle
J. Float Bowl Housing
K. Idle Chamber

FIGURE 68 Side Draft, Float Bowl Type
Carburetor

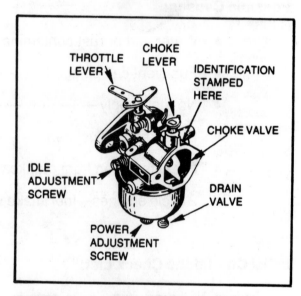

FIGURE 70 Type LMB Carburetor

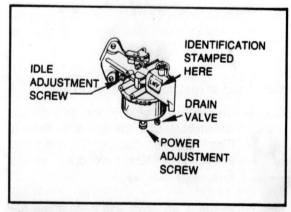

FIGURE 71 Type LMV Carburetor

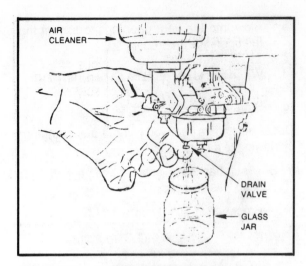

FIGURE 72   Draining Fuel into Glass Jar

e.   The fuel (gasoline or gasoline oil mixture) will float on top of any water and a distinct separation between water and fuel will be evident, if there is water in the fuel.

Is there any evidence of water in the fuel?

YES—Fault has been isolated: Water in fuel. Drain tank completely; refill with fresh fuel.

NO—Proceed to Step 2.

Step 2—Examine the fuel for dirt or rust contamination. This must be eliminated by cleaning out the fuel tank as follows:

a.   If there is any evidence of dirt, etc. in the fuel sample, completely drain tank into the jar and discard this fuel.

b.   Next, disconnect the fuel line going from the tank into the carburetor and permit this fuel supply line to drain free.

c.   Finally, flush out both the tank and supply line with fresh fuel. A final flush using a can of "DRY GAS," which may be purchased at any gasoline station or auto supply store, is recommended to completely eliminate any residual water in the tank or line.

d.   Permit the fuel tank and supply line to dry out about ½ to 1 hour. In the meantime dry off the spark plug, clean and regap the plug per Service Procedure **4004**.

Replace the spark plug, connect ignition cable, connect fuel supply line, close all drain valves and fill tank with clean fresh fuel. Start up engine.

Did the engine start up easily?

YES—Proceed to Step 3.

NO—After several tries, if engine does not start, refer to Fault Symptom **2000**.

Step 3—Permit the engine to run for about 10 to 15 minutes.

Does the engine run satisfactorily without any sputtering?

YES—Fault has been verified: Water or rust contamination in fuel.

NO—Cause not determined. Proceed to Fault Symptom **3120**.

3120

Fuel vent cap plugged

Step 1—This step provides a determination of whether or not the fuel tank vent is clogged. The air vent is usually incorporated into the design of the gas tank cap. If the vent is clogged, the air supply into the tank will be cut off, and a partial vacuum will develop inside the tank. This will shut off the flow of fuel to the engine. Proceed as follows:

a.   Place the THROTTLE control to OFF.

b.   Be sure the gas cap is finger tight.

c.   Start up the engine.

d.   Run engine at mid-speed.

e.   If the engine will not start slightly loosen the gas cap. Take particular notice as the cap is loosened of whether there is a whoosh or hissing sound made by the air.

Was there a hissing sound made by the air as the cap was loosened?

YES—Fault has been verified: Vent cap is plugged. Proceed to Step 3 to clean cap or replace with new cap.

NO—Proceed to Step 2.

Step 2—After the engine starts up, run the engine at mid-speed. Be careful that the gas cap does not loosen completely and vibrate off of the tank. It may be necessary to hold down the cap onto the tank. Use caution and stay clear of mower blade(s) or rotating parts. Do not have the cap so loose that fuel can spill on the engine.

If the engine runs well for about 3 minutes or so, tighten the gas cap. Permit the engine to run for 5 to 10 minutes longer.

Did the engine sputter or slow down and stop?

YES—Fault has been verified: Vent cap is plugged. Proceed to Step 3 to clean cap or replace with new cap.

NO—Proceed to Fault Symptom **3130**.

Step 3—Obtain a can of carburetor cleaner fluid (at any auto supply shop) and proceed as follows:

a. Place the gas cap in a small clean can, and pour the cleaner over the cap—covering it completely.

b. Let the cap soak in the cleaner for about 10 minutes.

CAUTION

USE MAXIMUM/AMPLE VENTILATION WHEN USING CARBURETOR CLEANER. WORK OUTDOORS PREFERABLY, AND FOLLOW DIRECTIONS OF CLEANER MANUFACTURER—FUMES ARE FLAMMABLE AND HARMFUL.

CAUTION
---

c. Remove the cap from the cleaner and shake off the cleaner fluid.

d. Wipe dry. Let the cap dry completely.

e. Blow into the vent openings to assure that the openings are free of dirt.

f. When cap is dry replace on tank.

g. Start up engine. Let engine run.

Does the engine run well without any sputtering or slowing to a stop?

YES—Fault has been isolated and verified: Vent cap was plugged.

NO—Proceed to Fault Symptom **3130**.

3130

No fuel supply

Step 1—When engine has difficulty running, sputters, slows and finally stops, check the fuel supply into the cylinder. Proceed as follows:

a. Place the THROTTLE to OFF.

b. Crank the engine a few times.

c. Remove the ignition cable and remove the spark plug (per Service Procedure **4004**).

Is the spark plug wet with fuel?

YES—Proceed to Fault Symptom **3140**.

NO—Proceed to Step 2.

Step 2—For mowers which use a fuel pump to supply fuel to the carburetor, the fuel pump may be faulty. For engines which have the fuel pump readily accessible and with a line running from the pump to the carburetor as shown in a typical installation of Fig. 73, test the pump's proper operation as follows:

Engines which do not have a fuel supply line going from the pump to carburetor require removal for examination of the pump. See Service Procedure **4020**.

a. Hold a pan below the fuel supply line which runs from the pump to the carburetor and then disconnect this line.

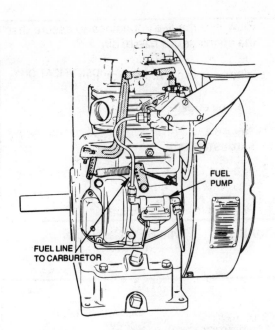

FIGURE 73 Typical Fuel Pump Installation for External Accessible Pump

b. Check that this line is free of dirt and gum deposits. If dirty, clean with carburetor cleaner.

c. Hold the pan just at or below the point where the line came out of the fuel pump, the outlet side of the pump.

d. Crank over the engine several times. The fuel pump should pump fuel freely out of the outlet into the pan.

Does the pump provide a freely running supply of fuel?

YES—Faulty pump is not verified. Proceed to Step 3.

NO—Fault has been isolated: Faulty or dirty fuel pump. Refer to Service Procedure **4020.**

Step 3—Hook up the line to the fuel pump but leave free at carburetor end. Crank engine as in Step 2 and check free flow of fuel out of the end of the line.

Does fuel run out freely?

YES—Fault has not been isolated. Proceed to Fault Symptom **3140.**

NO—Fault has been isolated to a clogged line. Re-examine and clean as per Step 2.

a. Reconnect lines and attempt engine start.

Did engine start easily?

YES—Faulty clogged fuel line.

NO—Fault not isolated. Proceed to Fault Symptom **3140.**

3140

Carburetor adjustment

Step 1—There are two primary adjusting devices on the carburetor which if improperly adjusted will cause the engine not to run properly. In addition if they are dirty or plugged up this adds to operating problems. These two adjusting devices for carburetor operation are the main fuel adjustment screw (also called power adjustment needle, or high speed adjustment needle) and the idle fuel flow adjustment screw (also called idle speed adjustment or idle speed adjustment screw, or idle speed regulating screw). These parts are identified on the carburetors shown in Figs. 74 to 80. Adjust these screws as follows:

A. Throttle Plate and Throttle Shaft Assembly
B. Idle Speed Regulating Screw
C. Idle Fuel Regulating Screw
D. Choke Plate and Choke Shaft Assembly
E. Atmospheric Vent Hole
F. Float Bowl Housing Drain Valve
G. Float Bowl Housing Retainer Screw
H. High-speed Adjusting Needle
J. Float Bowl Housing
K. Idle Chamber

FIGURE 74 Side Draft, Float Bowl Type Carburetor

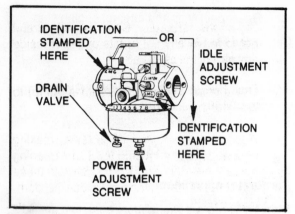

FIGURE 75   Type LMG Carburetor

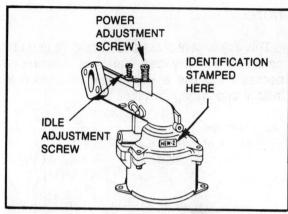

FIGURE 78   Type HEW Carburetor

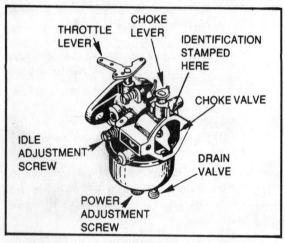

FIGURE 76   Type LMB Carburetor

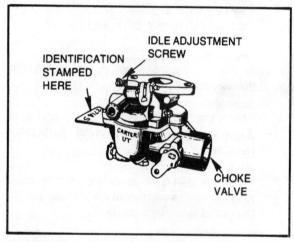

FIGURE 79   Type UT Carburetor

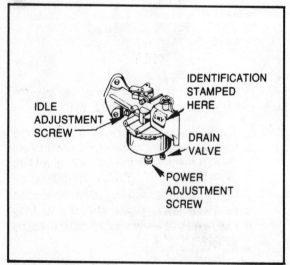

FIGURE 77   Type LMV Carburetor

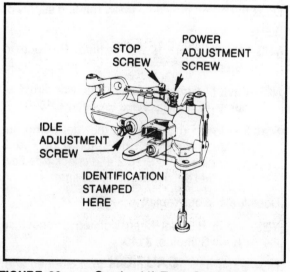

FIGURE 80   Suction Lift Type Carburetor

NOTE:

This adjustment procedure should be done by hand without any tool, however it may be necessary to use a screwdriver or wrench to initially loosen the adjustment screw.

a. Remove both the main fuel adjustment screw and the idle speed adjustment screw. Note carefully from where each screw is removed and tag screws if necessary.

b. Place screws in a pan and apply cleaner to screws to thoroughly degrease and clean the screws.

c. Pay particular attention to the screw tips to insure they are free of all gum and dirt.

d. Dry off the screws.

e. Remove the air filter assembly from carburetor.

f. Apply the cleaner down into the throat of the carburetor, and into the openings into which the two adjusting screws fit.

g. Apply the cleaner liberally so that the cleaner runs freely out of the screw openings.

```
┌─────────── CAUTION ───────────┐
│                               │
│                               │
│        USE CARBURETOR         │
│        CLEANER ONLY ON        │
│         A COLD ENGINE.        │
│       INSURE MAXIMUM          │
│       VENTILATION. PRE-       │
│       FERABLY USE OUT-        │
│        DOORS—SINCE            │
│        CLEANER FUMES          │
│         ARE VOLATILE          │
│         AND HARMFUL.          │
│        DRY ENGINE OFF         │
│        BEFORE STARTING.       │
│                               │
│                               │
└─────────── CAUTION ───────────┘
```

h. Dry off the carburetor. Be extremely careful not to leave any lint or bits of paper inside any part of the carburetor.

i. Use a small air pump (basketball pump) to blow out the carburetor.

j. Replace the adjustment screws, making sure to put each one into the same opening it was removed from. Do not force in these adjusting screws, since they can be damaged easily!

k. Using only your fingers—carefully screw the main fuel adjustment screw in until it is fully seated and finger tight. After it is seated, back the screw out 1 to 1½ turns.

l. In the same manner, screw the idle fuel adjustment in until it is fully seated—then back it out 1 turn.

m. Replace the air filter assembly on the carburetor, replace spark plug, ignition cable.

Start up engine and let the engine run 5 to 10 minutes, or longer. After this time period (or during this time) did the engine sputter or slow and stop?

YES—Fault is not verified. Proceed to Fault Symptom **3150**.

NO—Fault has been isolated: Faulty carburetor adjustment.

3150

Fuel starvation from crankcase air leaks—rotary mowers with vertical crankshaft

Step 1—A crankcase air leak will thin out the fuel-air mixture in the engine crankcase, and starve the engine for fuel, as well as raise the engine operating temperature to an undesirable level. Bad crankcase oil seals, leaking gaskets, or a leaking reed plate are the cause of crankcase air leaks.

For rotary mowers which have the engine mounted vertically (with the power end of the crankshaft vertical and the mower blade attached to the end of the crankshaft) proceed as follows:

a. Place the THROTTLE control to OFF or STOP, and disconnect the spark-plug ignition cable.

b. Tilt the mower up—being careful not to have the fuel tank or oil port spill over.

c. Examine the cutting blade and the cutting blade adapter (see Fig. 81). Look for evidence of raw unburned fuel on these parts. In normal operation these parts should have a dry gray to black exhaust gas film covering them. If the crankcase seal is leaky, then it is most probable that there will be indications of raw fuel on the blade and/or adapter.

Is there evidence of raw fuel on the blade and/or adapter?

YES—Fault has been isolated: Engine crankcase seal is leaking. Refer to Service Procedure **4100.**

NO—Proceed to Fault Symptom **3160.**

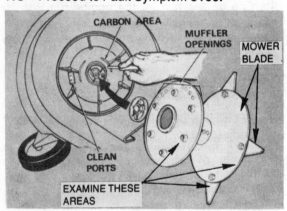

FIGURE 81   Examination of Cutting Blade Areas for Raw Unburned fuel

3151

Fuel starvation from crankcase air leaks—reel mowers and rotary mowers with horizontal crankshaft

Step 1—For rotary mowers which have the engine mounted horizontally (with the power end of the crankshaft horizontal) and reel mowers test leaky crankcase oil seals as follows:

a. If the engine has been running, wait until it cools before restart.

b. Be sure the pre-start checks of Fault Symptoms **1000** and **2000** have been followed. Make certain the CHOKE control does set the carburetor butterfly valve to full Choke.

c. Reel mowers, and mowers which have the engine crankshaft mounted horizontally have the crankcase seal on the side where the engine power takeoff is located. The power takeoff usually drives a clutch, a gear reduction box, a belt pulley, or a sprocket/-chair pulley. Start engine.

d. Run the engine with as slow an RPM or speed as possible.

CAUTION

BE ABSOLUTELY CERTAIN NOT TO ENGAGE THE CLUTCH OR POWER TAKEOFF. BE EXTREMELY CAREFUL WITH HANDS OR LOOSE GARMENTS, SO AS NOT TO GET CAUGHT IN ANY MECHANISM. IF IT IS NECESSARY TO REMOVE ANY BELT OR CLUTCH GUARD TO GET AT THE CRANKCASE SEAL, USE A LONG SPOUT OIL CAN, OR A LONG WOODEN ARTIST'S BRUSH TO PUT OIL AT SEAL.

CAUTION

FIGURE 82   Typical Location of Power Take-off

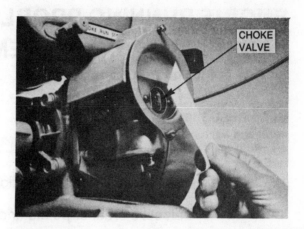

FIGURE 83   Test for Leaky Reed Valve

e.   Refer to Fig. 82. Squirt or place a small amount of engine oil around the crankcase seal.

f.   Listen carefully to the sound of the engine. Does the engine suddenly change the sound it made before you squirted the oil on the seal (to a smoother higher pitch sound after you oiled the seal)?

YES—Fault has been isolated: Leaky crankcase seal. Refer to Service Procedure **4100**.

NO—Proceed to Step 2.

Step 2—Look at the exhaust muffler when you squirt the oil on the seal. Does the exhaust smoke suddenly increase in volume and give off a heavy white smoke?

YES—Fault has been isolated: Leaky crankcase seal. Refer to Service Procedure **4100**.

NO—Proceed to Fault Symptom **3160**.

3160

2-cycle engines—fuel intake valve defects

Step 1—A leaky reed valve on 2-cycle engines will cause engine sputter and stalling. Test the reed valve as follows:

a.   Place controls or THROTTLE at OFF or STOP.

b.   Remove the carburetor air cleaner assembly.

c.   Start the engine. Follow Start-up Procedure **1000**.

d.   With the engine running hold a 1½ to 2 inch wide piece of white clean paper about 1 inch from the intake of the carburetor as shown in Fig. 83.

e.   Examine the paper.

Has the paper been spotted by the fuel mixture?

YES—Fault has been isolated: Leaky or bad reed valve. Refer to Service Procedure **4012**.

NO—Refer back to Fault Symptom **3000** for other fuel system tests on engine running problems.

ENGINE RUNNING PROBLEMS
FAULTY IGNITION SYSTEM   FAULT SYMPTOM   3200

Possible Causes:

- Defective spark plug ————————————————— **See 3210**

- Defective high tension ignition cable ——————— **See 3220**

- Defective breaker points or defective electrical system — **See 3230**

Initial Conditions Check List:

Perform all the pre-start checks and conditions given in Fault Symptoms **1000** and **2000**.

3210

Defective spark plug

Step 1—Position THROTTLE control to OFF or STOP. Remove ignition cable from spark plug, and remove plug per Service Procedure **4004**. Examine the electrodes and insulator tip of the plug. See Fig. 84 for nomenclature. Brown to gray-tan deposits on the insulator show a normal plug.

Does the spark plug look like the NORMAL plug shown in Fig. 85? Refer to Fault Symptom **2220** for greater detail on abnormal plug condition.

YES—Proceed to Fault Symptom **3220**.

NO—Fault has been isolated: Spark plug is dirty or fouled. Refer to Service Procedure **4004**.

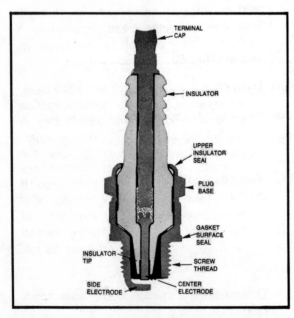

FIGURE 84   Spark Plug Nomenclature

FIGURE 85   Normal Plug

3220

Defective high tension ignition cable

Step 1—Engine running problems such as stalling, sputtering and rough running can be caused by the high voltage or high tension ignition cable being short circuited to ground. This prevents the high voltage from reaching the spark plug. Test the high tension cable for defects as follows:

a. Place controls at OFF or STOP.

b. Begin at spark-plug end of ignition cable and visually check all around the cable for its total length.

c. Look for cuts, cracks, worn spots in the cable insulation and where the cable may be worn so thin that the electrical conductor on the inside of the cable can be seen.

d. Pay particular attention to the points where the ignition cable passes over, or comes into contact with metal surfaces or corners.

e. If at a corner crossing the cable is worn thin, or the electrical conductor actually is in contact with a metal corner, then the cable is short-circuiting the electricity.

f. If in doubt run this test at night and crank the engine using the starter. A short circuit will be easier to see when the area around the cable is dark.

Are there any bare spots or metal contact points where the ignition cable is short-circuited?

YES—Proceed to Step 2.

NO—Proceed to Step 3.

Step 2—If a defective spot or point is found on the cable, temporarily repair that point. Wrap electrician's tape (available at any hardware store) around that point. The electrician's tape will act as an insulator and prevent the cable from becoming grounded.

When this temporary repair is complete, connect up the ignition cable to the spark plug. Start up the engine and let run for 5 to 10 minutes, or longer.

Does the engine run satisfactorily without stalling, sputtering or rough running?

YES—Fault has been verified: High voltage ignition cable was shorting to ground. Remember to make a permanent repair.

NO—Proceed to Step 3.

Step 3—The procedure of Step 2 provides inspection of the cable up to the point where the cable enters under the engine magneto flywheel or head, but no further. It is possible the cable may be defective beyond that point and even where the cable is joined to the magneto coil. There is no easy method to examine the cable under the engine magneto flywheel. The engine flywheel must be removed, thereby exposing both the terminal portion of the cable as well as the magneto assembly. The steps for this examination are given in Service Procedure **4030**.

3230

Defective breaker points or electrical system

3231

Defective electrical system, poor spark

Step 1—In Fault Symptom **3210** and **3220** tests, the spark plug and high tension ignition cable were found to be satisfactory. A simple test can be performed to determine if the breaker, points and the magneto coil/condensor system are generating a consistently good spark. If the spark is intermittent, or not of a consistently good quality, then some of the ignition system components must be serviced or replaced. Proceed as follows:

a. Disconnect the ignition cable from spark plug.

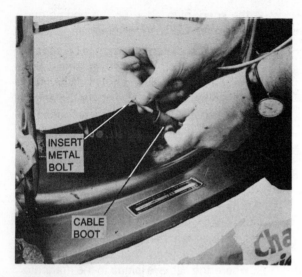

FIGURE 86   Insertion of Metal Bolt into Ignition Cable End Boot

b. Place the THROTTLE control at mid-position or RUN position.

c. If the end of ignition cable has a protective insulator boot, pull back the boot and expose cap. If this cannot be done, insert a small metal bolt into the cap so that the end of the screw is exposed as shown in Fig. 86.

d. Hold the end of the ignition cable by means of a wooden or plastic clothespin to avoid an electric shock.

e. Now hold the cable so that the end of the bolt—or the cable end cap—is about 1/8 inch away from the base of the spark plug. Simultaneously crank the engine. Be sure engine is cranked over at full speed.

f. A spark which is bright-blue in color should jump from the ignition cable cap or the bolt end to the base of the spark plug. The spark will make a snap or crackle sound as it jumps to the plug base.

Was there a snappy bluish-white spark?

YES—Proceed to Step 4.

NO—Proceed to Step 2.

Step 2—Was there any spark produced?

YES—Proceed to Step 3.

NO—Fault has been isolated: Defective ignition system. Refer to Service Procedure **4030** for correction of defect.

Step 3—Repeat Step 1 a number of times. Note carefully the color and intensity of the spark produced. If the spark color was not blue-white and it did not jump sharply, this indicates:

a. A poorly operating ignition/electrical generation system.

b. A spark which may not fire the spark plug consistently.

c. If in cranking the engine, or during any cranking period, a spark was missed—there was no spark—then this indicates the breaker points are worn or pitted and/or the magneto coil/condenser system are in need of overhaul.

Was the spark color yellow, orange, or yellowish orange?

YES—Proceed to Fault Symptom **3232.**

NO—If the spark fires intermittently, refer to Service Procedure **4030.**

Step 4—This step will determine if the basic magneto-electrical system is capable of delivering the required electrical voltage. Proceed as follows:

a. Disconnect the ignition cable from the spark plug.

b. Obtain a lawn mower spark plug (preferably a new plug, that is known to be good) and open the gap to between 5/32 to 3/16 of an inch as shown in the inset of Fig. 87. (An alternate test plug may be purchased at most lawn mower repair shops. This type of special test plug needs no further adjustments and should be purchased with an attaching clamp. See Fig. 88.)

c. Connect the ignition cable to the terminal of the test plug—and if equipped with an alligator attaching clamp, clamp to an engine bolt or a cooling fin so as to ground the thread portion of the test plug.

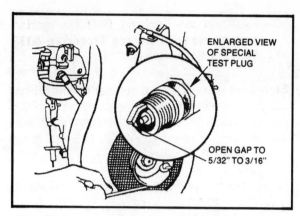

FIGURE 87   Wide Gap Test Plug

FIGURE 88   Special Test Plug

NOTE:

Be certain that the engine clamping point is a good ground point. The metal bolt head or fin should be bare and clean. Scrape away all paint or dirt before attaching clamp.

d.   If the test plug has no attaching clamp, use a test lead and clamp one end over the test plug thread and the other end to an engine bolt or fin. (Be sure to follow the clamping NOTE.)

e.   Place the THROTTLE control at RUN or at mid-position (turn ignition switch ON, if so equipped).

f.   Crank the engine and observe the electrode gap on the test plug. When the engine is properly cranked and with the ignition system in good condition the magneto-electrical system should be capable of delivering more than enough voltage, so that a spark will jump across the gap of the test plug.

Was there a spark at the test plug gap?

YES—The ignition system appears to be capable of delivering the required electrical voltage. The ignition timing may be varying or has slipped to a position where the engine is stalling out. Refer to Service Procedure **4030** for discussion and procedure for engine timing.

NO—Fault has been isolated: The ignition system is in need of overhaul or repair. Refer to Service Procedure **4030**.

3232

Defective breaker points

Step 1—This procedure applies to lawn mower engines that are equipped with ignition breaker (also called contact) points which do not require the removal of the magneto flywheel for points examination, or for replacement of the points. Fig. 89 shows the location of such ignition points on a Kohler lawn mower engine. For engines which do not have accessible points refer to Service Procedure **4030**.

The condition of the ignition breaker points are of prime importance for the proper operation of the engine. If the points are burned, oxidized or badly pitted, little or no electrical current will pass through them when they close. Fig. 90 shows some sketches of points in good condition or clean points, and also poor points which are badly pitted.

To examine and check accessible breaker points, proceed as follows (refer to Fig. 89):

a.   Place all controls to OFF or STOP.

b.   Remove the breaker points cover and cover gasket. Handle the gasket carefully so that it can be reinstalled.

c.   With a screwdriver, carefully move the end of the movable breaker arm away from the fixed arm so that you can see the flat contact surfaces of the points.

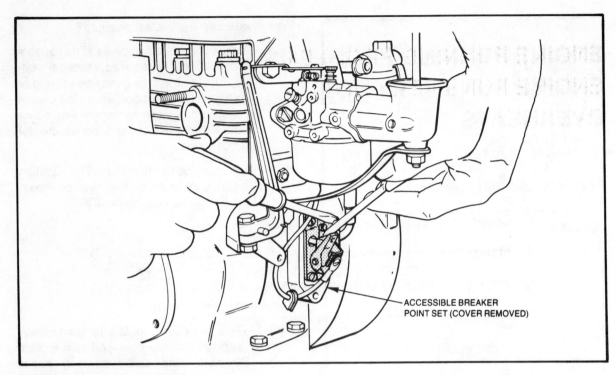

FIGURE 89 Typical Accessible Breaker Point Set

d. If the points are in good condition, they should look like "A" of Fig. 90. If the points are burned or pitted, they will require replacement. Refer to Service Procedure **4030**.

e. After visually examining the points and they appear to be in good condition, place the **THROTTLE** control at RUN or at mid-position (turn ignition switch ON, if so equipped). Remove the screwdriver.

f. Remove the ignition cable from the spark plug, and while cranking the engine over, look at the breaker points. If both the points and the ignition system are in good condition, a bluish-white spark will jump across the points when they open. This spark should occur consistently. The ignition should give a spark within the same time interval consistently.

Is there a spark across the points as the engine is cranked?

YES—This test has indicated that the points are in operating condition. Fault Symptom **3231** showed a poor quality spark was

produced but this spark may be marginal when the engine is running. Refer again to the maximum voltage output test in **3231**.

NO—Fault has been isolated: Ignition system breaker points require overhaul. Refer to Service Procedure **4030**.

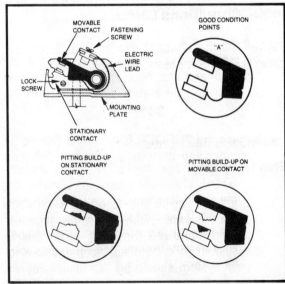

FIGURE 90 Good Condition Breaker Points and Poor Condition Pitted Points

ENGINE RUNNING PROBLEMS
ENGINE RUNS VERY HOT, OVERHEATS

FAULT SYMPTOM   3300

Possible Causes:

- Cooling fins obstructed/dirty, missing blower housing ———————————— See 3310
- Wrong fuel for 2-cycle engines ——————— See 3320
- Too lean fuel-to-air mixture, wrong carburetor settings ————————————————— See 3330
- Obstructed muffler ——————————— See 3340
- Obstructed exhaust ports on 2-cycle engines ——— See 3350
- Engine is continuously overloaded ————— See 3360
- Miscellaneous causes ———————————— See 3370

Initial Conditions Check List:

Perform all the pre-start checks and conditions given in Fault Symptoms **1000** and **2000**.

3310

Cooling fins obstructed/dirty

Step 1—Position the THROTTLE to OFF or STOP. Wait until engine cools down to the same temperature as the surrounding air. If the cooling system is faulty, or the cooling fins obstructed, the engine will run very hot and overheat. Proceed as follows:

a. Clean away all dirt, grass clippings or grease from all parts of the engine.

b. Be particularly careful to clean all of the fins around the engine cylinder. Dirt accumulations on the fins act as insulation and prevent the proper transfer of heat away from the engine.

Start up the engine and run for 10 to 15 minutes.

Does the engine run satisfactorily without any evidence of overheating?

YES—Fault has been isolated: Obstructed cooling fins.

NO—Proceed to Step 2.

Step 2—Check that shrouds, coverings and particularly blower housings have been kept intact on the engine and that they have not been removed and left off. If the blower housing has been removed and left off the engine, then the cooling fins are not circulating the air properly over the fins. This will cause the engine to overheat. In addition continued overheating will cause internal damage to the engine. This can render the engine useless.

Be certain all necessary shrouds, blower housing and the like are installed. Start up and run the engine 10 or 15 minutes.

Does the engine run well with no indication of overheating?

YES—Fault has been isolated: Engine overheating due to missing blower housing or shroud.

NO—Fault not isolated. Proceed to Fault Symptom **3320**.

3320

Wrong fuel for 2-cycle engines

Step 1—Proper operation of 2-cycle engines requires that the fuel mixture of oil and gasoline follow the engine manufacturer's specification exactly. See Table I for some typical manufacturer's mixture. Check the fuel mixture which has been used and if there is any doubt on the mixture, drain the fuel tank completely and refill. Be sure to follow the manufacturer's specification. Start up and run the engine for about 10 minutes. Is there any indication of the engine overheating?

YES—Fault not isolated. Proceed to Fault Symptom **3330**.

NO—Fault has been isolated: Wrong fuel mixture used.

3330

Too lean fuel-to-air mixture, wrong carburetor settings

Step 1—If the fuel-to-air mixture is set too lean, this will provide more air than required and cause the temperature inside the cylinder to increase. Proceed as follows to check for proper carburetor adjustments:

a. Shut down engine, place THROTTLE control at STOP. Allow engine to cool.

b. Refer to Figs. 91 to 97, for various makes of carburetors which are typically used on lawnmower engines. Two primary adjustment devices for proper carburetor operation are the main fuel adjustment screw (also called power adjustment needle, or high speed adjustment needle) and the idle fuel flow adjustment screw (also called idle speed adjustment, or idle speed adjustment screw, or idle speed regulating screw). These parts are identified on the carburetors shown in Figs. 91 to 97.

NOTE:

The adjustment procedure should be done by hand without any tool, however it may be necessary to use a screwdriver or wrench to initially loosen the adjustment screws.

c. Using only your fingers, carefully screw the main fuel adjustment screw in until it is fully seated and finger tight. DO NOT FORCE THIS SCREW IN SINCE THIS WILL DAMAGE THE SCREW.

d. After the screw is seated, back the screw out 1 to 1½ turns.

e. Perform the same operation for the idle screw adjustment. DO NOT FORCE THE SCREW. Turn the screw in until it is seated—then back out 1 turn. Proceed to Step 2.

Step 2—The settings for the power and idle screws as given up to this point have been general settings. In order to insure that the engine will not run too slow or too fast, it is necessary to fine tune or fine adjust these settings. Fine tuning requires that the RPM (the speed) of the engine be adjusted for both idle running and top speed running. This procedure will set the engine idle and high speed,

A. Throttle Plate and Throttle Shaft Assembly
B. Idle Speed Regulating Screw
C. Idle Fuel Regulating Screw
D. Choke Plate and Choke Shaft Assembly
E. Atmospheric Vent Hole
F. Float Bowl Housing Drain Valve
G. Float Bowl Housing Retainer Screw
H. High-speed Adjusting Needle
J. Float Bowl Housing
K. Idle Chamber

FIGURE 91   Side Draft, Float Bowl Type Carburetor

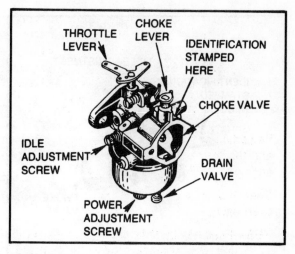

FIGURE 93   Type LMB Carburetor

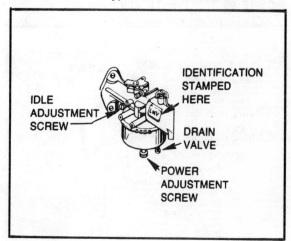

FIGURE 94   Type LMV Carburetor

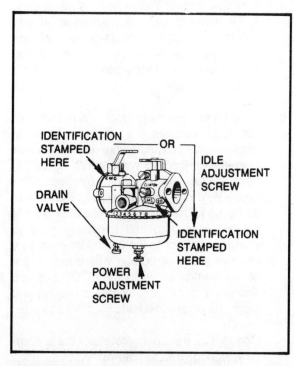

FIGURE 92   Type LMG Carburetor

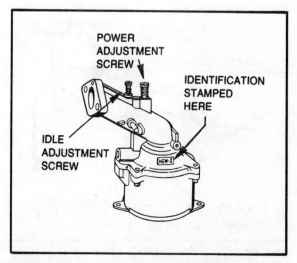

FIGURE 95   Type HEW Carburetor

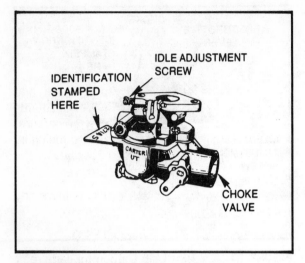

FIGURE 96   Type UT Carburetor

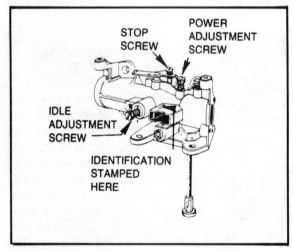

FIGURE 97   Suction Lift Type Carburetor

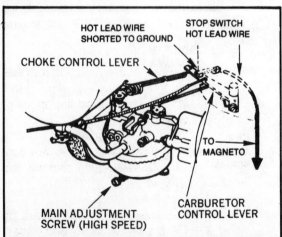

FIGURE 98   Typical Location of Hot Lead Stop Switch Wire-Engine Control in "Stop" Position

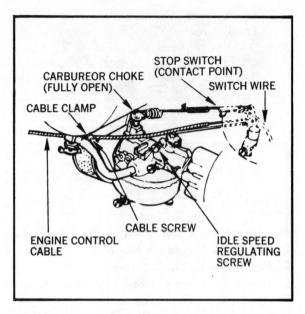

FIGURE 99   Typical Positions of Control Cable for Fast or High Speed Position

by means of a direct reading of the revolutions per minute (RPM) which the engine is running at.

a. A simple engine Tachometer is necessary. A Tachometer can be purchased at any automobile supply store, larger department store, and at most electronic supply shops. An inexpensive Tachometer is fully sufficient for this task.

b. It will be necessary to locate the hot point or hot lead wire on the engine. On most engines the hot lead wire is the smaller diameter of the two wires which come out from under the magneto flywheel. The heavy wire is the high tension ignition cable, this goes to the spark plug. The other wire is the hot lead wire and it usually goes to starter switch, or to the control lever, or to a point near the carburetor where movement of the control cable to the STOP position shorts the hot lead (voltage) to ground and stops the engine. Refer to Figs. 98 and 99.

c. Connect up the Tachometer per the accompanying Fig. 100, as follows:

•   The "hot point" is the point where the primary wire of the magneto coil is connected along with the condenser wire, and the breaker points. It is from this common point that the hot lead wire goes to the starter switch or control cable short-out location.

The black or ground lead to the Tachometer can be connected to any convenient ground point, but using the base of the spark plug is preferred.

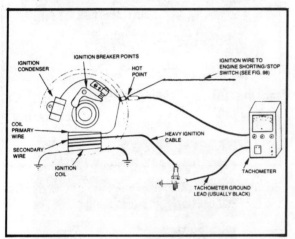

FIGURE 100   Tachometer Hook-up Connections

d. Refer to the engine manufacturer's specification for the idle speed RPM (or Service Procedure **4050**). If the information is not available, then use an idle RPM speed of 1300 or 1400 RPM. Top speed is about 3500 RPM. Proceed to adjust the speed as noted.

e. Start up engine, let engine warm up at IDLE control lever position about 5 minutes, meanwhile read the RPM on the Tachometer.

f. Using your fingers adjust the *IDLE* screw until the desired idle RPM is reached.

g. Reposition the control lever to maximum or high speed. Let engine run at this speed a few minutes.

h. Using your fingers adjust the *power* screw until the desired RPM is reached.

i. Leave the Tachometer connected, and move the control back and forth from idle to high speed. Note that the engine should respond smoothly and should stay at the

speed to which the control lever is positioned. The idle RPM and top speed RPM should be those speeds as adjusted by the carburetor screws.

j. Move the THROTTLE control back and forth a number of times to insure the speeds read out are those selected by the carburetor adjustments. If not, some fine tuning by small adjustments may be necessary.

After the engine has run for a while is there any evidence of overheating?

YES—Proceed to Fault Symptom **3340.**

NO—Fault has been isolated: Wrong carburetor settings.

3340

Obstructed muffler

Step 1—A problem that could cause the engine to run very hot is an obstructed exhaust muffler. For both 4-cycle engines and 2-cycle engines if the muffler is obstructed by dirt or carbon or if the muffler is of the wrong type, this will cause overheating. If the muffler has been replaced, check with the manufacturer's specification to insure an exact replacement was made. Check for an obstructed muffler as follows:

a. Position control lever to OFF, let engine cool.

b. Examine the open end of the muffler to see if foreign material such as dirt, grass clippings, etc. may have filled the opening. Remove these.

c. Make sure the engine and muffler are cool to the touch, then carefully remove the muffler from the engine.

d. Usually the muffler can be unscrewed by hand; if not, use a wrench at the end closest to the engine. Be careful not to damage the muffler. If the muffler does not move, then apply a few drops of penetrating oil or liquid wrench (which is a very light oil used on rusty parts to loosen them) to the threaded part going into the engine block. Wait a few minutes, for the oil to penetrate, and then unscrew the muffler.

e. Examine the intake side of the muffler for carbon clogging.

Is the muffler clogged or obstructed?

YES—Proceed to Step 2.

NO—Proceed to Fault Symptom **3350.**

Step 2—If the muffler is clogged, discard. Replace with the exact muffler. Internal cleaning of a clogged muffler is difficult or impossible.

When the muffler has been replaced, be sure it is installed tightly, then start up engine and let run for 5 to 10 minutes.

Is there any indication of the engine overheating?

YES—Proceed to Fault Symptom **3350.**

NO—Fault has been isolated and verified: Clogged muffler.

3350

Obstructed exhaust ports on 2-cycle engines

Step 1—Two-cycle engine exhaust ports are built into the walls of the cylinder as shown in the sketch of a cross-section of a 2-cycle engine, Fig. 101. Engines which are mounted horizontally, with the crankshaft horizontal, incorporate a muffler which is part of the cover over the exhaust ports. Engines with vertically mounted crankshafts are discussed in Step 3. Proceed to Step 2.

Step 2—These exhaust ports and muffler opening must be kept free of carbon deposits. Should carbon be permitted to build up at the exhaust ports to the point where the ports are becoming blocked, then the engine can overheat. See Figs. 102 and 103. Proceed as follows to clean out and/or check the exhaust ports:

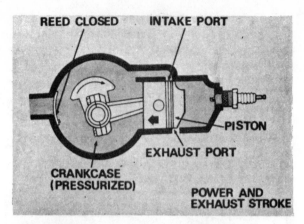

FIGURE 101   Typical 2-Cycle Engine Exhaust Ports

FIGURE 102   Check and Removal of Carbon Deposits on 2-Cycle Engine Exhaust Ports

FIGURE 103   Examination of Piston Rings after Carbon Removal from Exhaust Ports

┌─────────── **CAUTION** ───────────┐

IF ENGINE IS HOT
WAIT UNTIL ENGINE
COOLS BEFORE
PROCEEDING.

└─────────── **CAUTION** ───────────┘

a. Place controls or THROTTLE at OFF or STOP. Remove ignition cable from spark plug.

b. Remove the cover plate/or muffler over the engine exhaust ports. On horizontal mounted engines the exhaust ports/muffler will usually be on the side of the cylinder. Use care to select the correct wrench which fits the cover bolt heads and place the cover and bolts in a container for safekeeping.

c. Use a narrow paint scraper, or knife with a square blunt end blade, to scrape the carbon from the exhaust port openings.

┌─────────── **CAUTION** ───────────┐

BE CERTAIN THAT THE
CYLINDER PISTON IS
NOT RAISED UP TO
THE LEVEL OF THE
EXHAUST PORT HOLES
WHEN CLEANING
AROUND THE HOLES—
USE THE STARTER TO
LOWER THE PISTON
TO BELOW THE LOW-
EST PORT HOLE—
SCRAPING THE PISTON
WILL DAMAGE THE
PISTON.

└─────────── **CAUTION** ───────────┘

d. Replace the exhaust port cover and/or muffler. Be certain bolts are tight.

e. Reconnect the ignition cable to spark plug and start up the engine. Let run for about 10 minutes.

Is there any indication of the engine overheating?

YES—Proceed to Fault Symptom **3360.**

NO—Fault has been isolated: Blocked exhaust ports.

Step 3—On 2-cycle engines which are mounted vertically, with the crankshaft vertical, the exhaust ports are usually designed on the bottom or underside of the engine cylinder. This permits the exhaust gases to be deflected below the deck of the mower and onto a circular muffler. See Fig. 104. Proceed as follows to clean out and/or check the exhaust ports:

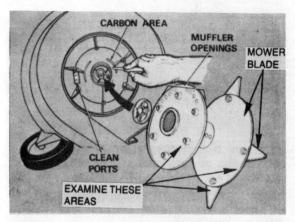

FIGURE 104   Typical 2-Cycle Engine, Vertically Mounted Crankshaft Exhaust Ports

┌─────────── **CAUTION** ───────────┐

IF ENGINE IS HOT
WAIT UNTIL IT
COOLS BEFORE PRO-
CEEDING.

└─────────── **CAUTION** ───────────┘

a. Place controls on THROTTLE at STOP or OFF. Remove ignition cable from spark plug.

b. Tilt the mower up so as to gain access to the bottom of the deck. Be careful not to have the fuel tank or oil port spill over. Prop up the mower with blocks of wood.

c. Remove the lawn mower blade (refer to Service Procedure **4002**).

d. Remove the bolts which are holding the muffler, place bolts in a container for safekeeping.

e. Remove the exhaust port cover, if there is one.

f. Use a narrow paint scraper, or knife with a square blunt end, to scrape the carbon from the exhaust port openings.

CAUTION

BE CERTAIN THAT THE CYLINDER PISTON IS NOT RAISED UP TO THE LEVEL OF THE EXHAUST PORT HOLES WHEN CLEANING AROUND THE HOLES—USE THE STARTER TO LOWER THE PISTON TO BE- LOW THE LOWEST PORT HOLE— SCRAPING THE PISTON WILL DAMAGE THE PISTON.

CAUTION

---

g. After cleaning out the carbon from the exhaust ports, clean out any carbon deposits from the mower deck. Carbon will tend to accumulate on and around any ridge or depression in the deck. The deck area should be scraped free of all carbon.

h. Replace the exhaust port cover and/or muffler. Be certain bolts are tight and be certain to include any lock washers which were under the bolt heads. Replace the cutter blade and be sure the blade bolts are tight. Place mower level.

i. Reconnect the ignition cable to the spark plug and start up the engine. Permit the engine to run for about 10 minutes.

Is there any indication of the engine overheating?

YES—Proceed to Fault Symptom **3360**.

NO—Fault has been isolated: blocked exhaust ports.

3360

Engine is continuously overloaded

When there appears to be no obvious reason for the engine to overheat, then it can be suspected the engine is being continuously overloaded. The amount of load and the manner in which the load is applied must be determined so that the mower horsepower can handle the load. It is possible that the mower engine is not of a sufficiently high enough horsepower. Consult your local lawn-mower dealer on the size engine recommended for the particular mowing load you have.

3370

Miscellaneous causes

Step 1—There are a number of miscellaneous causes of overheating. These are:

a. Heavy carbon deposits inside the engine cylinder. If carbon deposits have built up over a time period, these deposits serve as an insulation and prevent the cylinder from properly transferring heat to the fins and surrounding air. Removal of these deposits is a major overhaul process. Refer to Service Procedure **4100**.

b. Ignition system is not properly timed. Refer to Service Procedure **4030**.

c. Tappet clearance (on 4-cycle engines) set too tight. Refer to Service Procedure **4100**.

ENGINE RUNNING PROBLEMS
ENGINE LACKS POWER

FAULT SYMPTOM   3400

Possible Causes:

- Faulty fuel system ——————————— **See 3401**

- Incorrect control position settings ————— **See 3402**

- Faulty exhaust system ———————— **See 3403**

- Faulty ignition system ————————— **See 3404**

- Faulty valves or seals ————————— **See 3405**

- Faulty governor setting ————————— **See 3406**

- Poor mechanical condition, worn piston rings ——— **See 3407**

Initial Conditions Check List:

a. Perform all the pre-start checks given in Fault Symptom **1000** and Fault Symptom **2000**.

b. A common problem which occurs with lawn mowers is the mower begins to lose power. This can occur due to the normal process of wear. In many cases faults develop in the fuel system, or the control settings can slip or loosen out of adjustment, etc. The exhaust system, ignition, or valves may also develop problems. These problems will keep the engine from developing full power.

Additionally, if the lawn mower engine appears to be laboring, by working harder and harder, and then slows down in speed in the process of driving the cutting blade(s), the result will be a poorly cut lawn. The grass will be cut unevenly, clumps of grass may be torn out because the engine and blade(s) speed is not uniform.

In the discussions which are given herein, the major possible causes of loss of engine power are noted. When a specific fault is isolated the reader is directed to more detailed discussion of that fault and specific remedies.

3401

Faulty fuel system

Step 1—Start up and run the engine. Position THROTTLE at RUN, or mid-position of speed. Then move the throttle back and forth about the mid-position.

Does the engine speed respond to the throttle movement?

YES—Fault is not verified. Proceed to Fault Symptom **3402**.

NO—Fault has been verified. Proceed to Fault Symptom **3410** for fuel system tests.

3402

Incorrect control position settings

Step 1—Position the THROTTLE control lever to OFF. On many mowers moving the control lever from OFF or STOP to START, provides choking of the carburetor for start up, and fast to slow speed control. Therefore the correct installation of the control cable is important and may be the reason for lack of power. In the course of operation of the mower the control cable's actual positions on the carburetor may have changed so that the positions are not the same as are marked on the mower control panel or handle. Proceed as follows to insure that the control panel/handle position markings correspond to the proper carburetor lever positions:

a. Refer to Fig. 105 and 106 for sketches of a typical engine control cable connection wherein the STOP position and the FAST position are shown. Particular note should be taken of the fact that this flexible control cable (in this instance—other mowers may use more than a single cable) when at STOP position:

- Closes off the carburetor choke.
- Makes contact with the ignition ground contact point.
- Closes off the throttle plate (also called throttle valve or throttle butterfly).

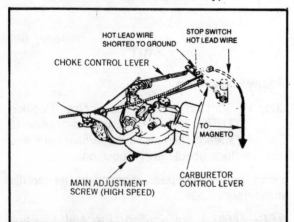

FIGURE 105   Typical Location of Hot Lead Stop Switch Wire-Engine Control in "Stop" Position

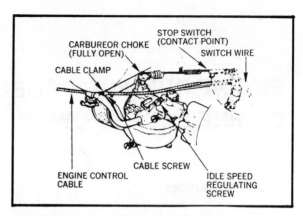

FIGURE 106   Typical Positions of Control Cable for Fast or High Speed Position

b. Examine the action of the mower control cable, when placed at the STOP position.

Does the end of the control cable accomplish all of the items listed under "a" above?

YES—Proceed to Step 3.

NO—Fault has been isolated: Incorrect control position setting. Proceed to Step 2.

Step 2—It will be necessary to adjust the control cable. Make a simple sketch or note of the markings on the control panel or handle. Position the control cable to STOP and examine the position of the cable and levers at the carburetor. Add these positions to your simple sketch. The choke control lever should be positioning the choke valve (or butterfly) so that it is fully closed. Compare this with your sketch information. The ignition contact point should be in good contact with the stop switch. Again compare this with your sketch information. If necessary, remove the air filter assembly from the carburetor to positively check that the choke butterfly is fully closed. Proceed as follows to adjust the control lever:

a. When the choke butterfly does not fully close adjust the control cable.

b. Loosen the cable clamp screw so that the cable flexible housing (the outer cable housing) can slide back and forth.

c. Move the control cable lever at the mower control panel, or handle, to the FAST or HIGH SPEED position. Check the following:

- The choke butterfly should be fully open.
- Look into the carburetor throat. The throttle butterfly should be fully open.
- If necessary, move the control lever slightly back and forth until a position is reached where the choke and throttle butterfly valves are fully open.

d. Tighten the cable clamp screw. Examine the control lever position on the control lever panel or handle. Note that the panel marking should agree with the FAST or HIGH setting. If the marking does not agree, it may be necessary to reposition the control panel. This should be done if the marking is too far off or out of agreement.

e. Move the control lever to STOP position. Check the cable position at the carburetor as follows:

- The ignition contact point should now be in solid contact with the stop switch.
- The choke butterfly valve should be fully closed (choked).
- If good contact is not made at the ignition stop switch adjust the fixed stop point or the lever for good contact.
- Be sure the contact areas are clean and free of paint, grease, oil or rust.

f. Move the control lever to the START (or CHOKE) position. Check that the choke butterfly valve is fully closed. Note that the throttle butterfly valve is now slightly open. If the choke is not fully closed, adjust the choke control lever or spring until the choke is fully closed.

g. Move the control lever back and forth a few times to insure that the lever positions agree with the carburetor settings and do not vary with movement of the control lever.

h. Proceed to Step 3.

Step 3—Start up the engine. The engine should start easily when the control lever is positioned at START. The engine should idle when the lever is positioned

at IDLE or SLOW. When the lever is at FAST or HIGH, the engine should perform accordingly. Movement of the control lever from slow to high should give a smooth response from the engine. A final adjustment by means of the main carburetor power screw and establishment of maximum RPM (speed) can be made by referring to Service Procedure 4010, Step 12.

Test the engine power and the adjusted control position setting by performing the mower cutting power test described in Fault Symptom 3030.

Does the mower perform the power cutting test satisfactorily?

YES—Fault has been isolated and verified: Incorrect control position settings.

NO—Proceed to Fault Symptom 3403.

3403

Faulty exhaust system

Step 1—Defects in the engine exhaust system will keep the mower engine from developing full power. These defects are a clogged or obstructed muffler or clogged exhaust ports. Two-cycle engines are particularly affected by clogged exhaust ports.

Fault Symptom 3050 discusses the various symptoms of a clogged muffler or clogged exhaust ports. The key symptom is that the engine becomes progressively quieter in operation (for 2-cycle engines) or shows increasing difficulty in cutting the grass.

Does the mower exhibit both of these symptoms?

YES—Fault has been isolated. Proceed to Fault Symptom 3420.

NO—Proceed to Fault Symptom 3404.

3404

Faulty ignition system

Step 1—When the ignition system begins to wear, less and less of the necessary voltage will be generated to fire the spark plug properly. This will eventually result in the engine failing to fire in each and every cycle. The mower engine will then begin to run less smoothly, show increasing difficulty in its ability to cut the grass, and finally stall very frequently.

Does the mower exhibit these symptoms?

YES—Fault has been isolated. Proceed to Fault Symptom **3430**.

NO—Proceed to Fault Symptom **3405**.

3405

Faulty valves or seals

Step 1—Defective intake or exhaust valves in 4-cycle engines will affect the mower engine power. Defective oil seals in both 4-cycle and 2-cycle engines will also affect the mower engine power. The end result of defects in valves or seals will be insufficient compression within the engine cylinder. When this happens the engine is no longer capable of developing the full power output it was designed for. Refer to the tests described in Fault Symptom **2550** compression readings and perform the tests.

Does the test result show poor compression readings?

YES—Fault has been isolated. Proceed to Fault Symptom **3440**.

NO—Proceed to Fault Symptom **3406**.

3406

Faulty governor setting

Step 1—A possible cause of lack of engine power is that the engine speed governor setting has changed or slipped from the factory set position. This can be due to vibration or a defect in the governor linkage.

Fault Symptoms **3401, 3402, 3403, 3404** and **3405** should all be reviewed and eliminated as possible causes of lack of engine power before proceeding further. Proceed to Step 2.

Step 2—Perform the following check on the engine governor:
a. Start up the engine.

b. After engine warm-up, position the THROTTLE control at HIGH, or FAST.

Does the engine speed stay at this high speed setting?

YES—Fault is not verified. Review the possible causes of lack of engine power **3400**.

NO—Fault has been isolated. Proceed to Fault Symptom **3450**.

3407

Poor mechanical condition, worn piston rings

Step 1—Internal wear of the engine cylinder walls, the piston rings, or both can result in lack of power. The engine cannot develop full compression if these defects exist. Refer to Fault Symptom **2505**, Steps 1 and 2. Perform the compression reading test cited in **2550**.

Does the engine show poor compression readings?

YES—Fault has been isolated. Proceed to Service Procedure **4100** for correction. Upon completion repeat Start-up Procedure 1000.

NO—Fault is not verified. Review the possible causes of lack of engine power **3400**.

ENGINE RUNNING PROBLEM
LACK OF POWER,
FUEL SYSTEM FAULTY

FAULT SYMPTOM   3410

Possible Causes:

- Wrong fuel (mixture) ————————————— **See 3411**

- Air cleaner clogged ————————————— **See 3412**

- Dirty fuel tank, dirty or restricted fuel line ———— **See 3413**

- Insufficient fuel supply from faulty fuel pump —— **See 3414**

- Incorrect carburetor adjustments ————— **See 3415**

Initial Conditions Check List:

Perform all the pre-start checks and conditions given in the Check List of Fault Symptom **2000**.

3411

Wrong fuel (mixture)

Step 1—Use of the wrong fuel, or wrong fuel mixture in the case of 2-cycle engines, can affect the engine's power output. If 2-cycle fuel is inadvertently used for a 4-cycle engine, the extra oil can speed up spark plug fouling, for example. See Table I for some typical manufacturer's fuel mixtures and note particularly Figs. 107, 108 on 2-cycle/4-cycle engines. Proceed to Step 2.

Step 2—Examine the container from which fuel is being used. Compare the fuel with the manufacturer's recommendations. For 2-cycle engines it is critical that the correct amount of oil be added in the mix. If there is any doubt on the fuel in the fuel tank, drain the tank completely. Refill with the manufacturer's recommended fuel. Proceed to Step 3.

Two-cycle—each cylinder fires once for each revolution. Air and fuel mixture is drawn into crankcase by piston on upward (compression) stroke. Exhaust stroke of piston compresses crankcase mixture, forcing it through by-pass into combustion chamber, where the incoming charge blows exhaust gasses out through exhaust port. Thus exhaust and intake cycle takes place in half a revolution of crankshaft.

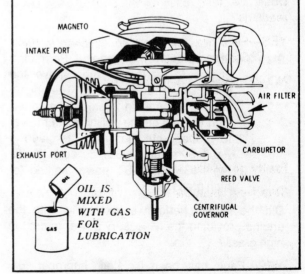

FIGURE 107   Two-Cycle Engine Cross Section View and Fuel Mixture

Four-cycle—each cylinder fires once for every two revolutions of crankshaft. All lubrication is supplied from oil in the crankcase that is either pumped or splashed to lubrication points. Poppet valves actuated by a crankshaft-driven cam open and close to permit the entry of intake charges and the expelling of exhaust charges. Downward stroke draws in charge of fuel and air; upward (compression) stroke prepares charge for firing and power stroke. The cycle is completed with the exhausting of the spent charge.

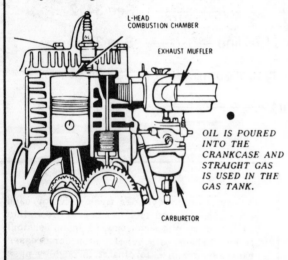

L-HEAD
COMBUSTION CHAMBER

EXHAUST MUFFLER

OIL IS POURED INTO THE CRANKCASE AND STRAIGHT GAS IS USED IN THE GAS TANK.

CARBURETOR

FIGURE 108   Four-Cycle Engine Cross Section View and Fuel-Oil Information

Step 3—Start up and run the engine. Permit the engine to warm up, then move the control level back and forth. The engine should respond in a smooth manner. Run the test on the ability of the mower to properly cut the grass, as given in Fault Symptom **3030**. Be sure to use the HIGH or FAST power setting for high grass.

Did the mower perform satisfactorily, with the engine providing the necessary power to cut the high grass?

YES—Fault has been isolated: Improper fuel (mixture).

NO—Proceed to Fault Symptom **3412**.

3412

Air cleaner clogged

Step 1—An inadequate supply of air will keep the engine from developing full power. The air cleaner must be maintained in proper condition since clogging, or a dirty air cleaner, can result in a very rich fuel-to-air mixture. This will cause inefficient engine operation. Shut down the engine, and position the THROTTLE to OFF. Proceed as follows to service the air cleaner:

a.   Remove the air cleaner and filter element and if the element can be cleaned, service it as shown in Figs. 109 to 113.

b.   Follow the procedure given in the figures for adding a light film of oil to the filter element.

c.   Clean out the filter container.

d.   Reinstall the element (or replace as required) and replace the air cleaner on the engine (if it was necessary to remove the air cleaner container).

Start up and run the engine. Permit warm-up, and then repeat the power test outlined in Fault Symptom **3030** on high grass, using HIGH power setting.

Did the mower perform satisfactorily, with the engine providing the necessary power to cut the high grass?

YES—Fault has been isolated: Inadequate supply of air into engine-clogged air filter.

NO—Proceed to Fault Symptom **3413**.

3413

Dirty fuel tank, dirty or restricted fuel line

Step 1—When the fuel flow to the carburetor is reduced or restricted, the engine will not be able to develop full power. The following elements of the fuel supply system must be checked to determine if the system is operating properly. Proceed to Step 2.

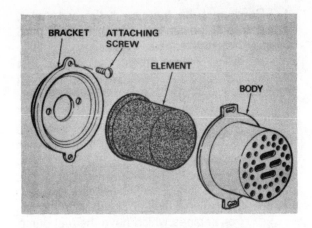

FIGURE 109   Removal of Air Filter

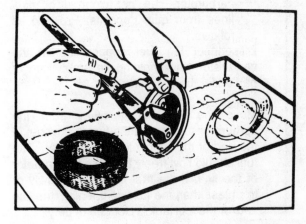

FIGURE 112   Oil Bath Type Air Cleaner

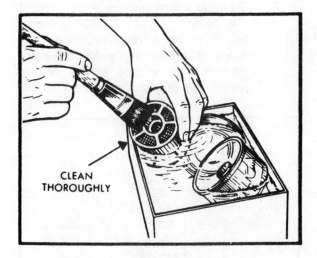

FIGURE 110   Cleaning Metallic Filter

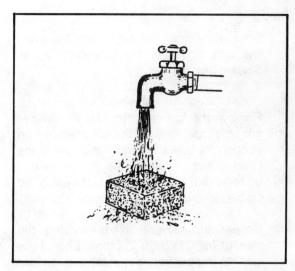

FIGURE 113   Polyurethene Element Type Air Cleaner

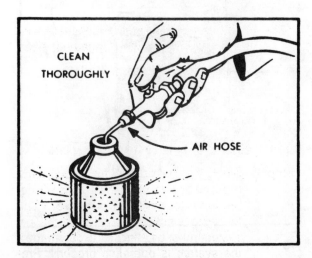

FIGURE 111   Dry Paper Type Filter

Step 2—Check for a dirty fuel supply line or a dirty fuel tank. Mowers with gravity feed of fuel to the carburetor, or mowers using a fuel pump to supply the carburetor, can get clogged or dirty lines from gum deposits. Proceed as follows:

a. Disconnect the fuel supply line at the carburetor while holding an empty can or pan under the line fitting.

b. Move the line slightly away from the carburetor and permit the fuel to flow out of the end of the supply line into the can. The fuel should flow freely from the full diameter of the line. If the fuel flows very slowly or from less than the full line opening, the line and/or the fuel tank is clogged or dirty.

Is fuel flow adequate?

YES—Fault not isolated. Proceed to Step 4.

NO—Proceed to Step 3.

Step 3—

a. Completely drain the tank from the open line and also from the tank drain valve. Discard this fuel.

b. Obtain a can of carburetor cleaner fluid. Remove the fuel cap and pour the cleaner into the fuel tank. Let the cleaner run through the tank and into a pan or can at the open tank drain valve. If the drain valve can be removed, remove it and clean using the carburetor cleaner.

c. Repeat this several times, cycling the cleaner back through the tank. Should the cleaner appear very dirty, use a fresh can for the last cleaning.

d. Examine the inside of the tank. If all the dirt and gum has been cleaned out, replace the drain valve and close the valve.

e. Pour a small amount of fuel in the tank. Be sure it runs freely out of the end of the open line into a can. If not, repeat the cleaning process until the line is completely open and the fuel flows freely.

f. Reconnect the fuel line to the carburetor, and refill the tank with fresh fuel. Start up and run the engine. Permit the engine to warm up, then repeat the power test outlined in Fault Symptom **3030** for high grass. Use HIGH power control setting.

Did the mower perform satisfactorily?

YES—Fault has been isolated: Dirty fuel tank or dirty, clogged fuel line.

NO—Proceed to Step 4.

Step 4—Check for a dirty or clogged fuel pump. For engines which have the fuel pump readily accessible, and with a line running from the pump to the carburetor as shown in a typical installation of Fig. 114, the following steps afford a test of the pump's proper operation [Engines which do not have a fuel supply line going from the pump to carburetor require removal and service of the pump. See Service Procedure **4020**.]:

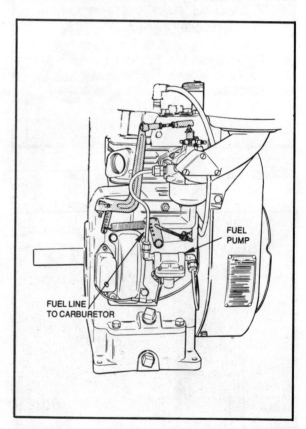

FUEL PUMP

FUEL LINE TO CARBURETOR

FIGURE 114   Typical Fuel Pump Installation for External Accessible Pump

a. Be sure that the ignition cable has been disconnected from the spark plug, and the control lever is at OFF, or STOP.

b. Hold a pan below the fuel supply line which runs from the pump to the carburetor and then disconnect this line.

c. Hold the pan just at or below the point where the line came out of the fuel pump, the outlet side of the pump.

d. Crank over the engine several times. The fuel pump should pump fuel freely out of the opening into the pan.

Does the fuel flow freely out of the fuel pump?

YES—Fault is not verified. Proceed to Fault Symptom **3415**.

NO—The fuel pump is clogged or dirty. Proceed to Step 5.

Step 5—Reconnect the fuel supply lines and tighten any fitting which may have been loosened. Reconnect the ignition cable and proceed as follows:

a. Obtain a can of carburetor cleaner fluid and add to the fuel in the fuel tank as shown in Fig. 115.

FIGURE 115   Addition of Carburetor Cleaner to Fuel in Fuel Tank

NOTE:

For 2-cycle engines be sure to allow for use of the cleaner fluid in place of gasoline in the gasoline-oil mix.

b. Follow the carburetor cleaner manufacturer's directions for adding the cleaner.

c. Start up and run the engine. Permit engine warm-up.

Repeat the power test outlined in Fault Symptom **3030** for high grass. Use the HIGH or FAST control setting.

Did the mower perform satisfactorily?

YES—Fault has been isolated: Clogged or dirty fuel pump.

NO—Proceed to Fault Symptom **3414**.

3414

Insufficient fuel supply from faulty fuel pump

Step 1—Run the engine after completion of Step 4 of **3413** where carburetor cleaner was added to the fuel tank. Run the engine until the mixture of fuel and cleaner is almost used up. Then shut down the engine, place control to OFF or STOP and proceed as follows:

a. Wait until engine cools.

b. Disconnect ignition cable.

c. Disconnect the fuel line at the outlet side of fuel pump as per Step 4 of **3413**.

d. Crank over the engine per Step 4 of **3413** while holding a pan at outlet side of pump.

Does the fuel pump supply fuel freely out of the pump while the engine is cranked?

YES—The fault is not verified. Proceed to Fault Symptom **3415**.

NO—The fault has been isolated: Faulty fuel pump. Refer to Service Procedure **4020** for removal and service of the fuel pump. Upon reinstallation of a properly operating fuel pump perform the grass cutting test given in Fault Symptom **3030** to verify that the fault has been eliminated.

3415

Incorrect carburetor adjustments

Step 1—When the carburetor settings for correct air-fuel and fuel quantity are improperly adjusted the engine will not develop full power.

In addition, on carburetors which incorporate an internal float and float needle valve, too low a setting of the float will prevent the engine from developing full power. The adjustments will differ depending on the type and manufacturer of the carburetor.

Refer to Fault Symptom **2170,** Step 2, and carry out the removal, cleaning, reinstallation and adjustment of the carburetor idle fuel flow and power adjustment valves.

Upon completion of this **2170,** Step 2, procedure, start up and run the engine. Permit engine warm-up, then repeat the power test outlined in Fault Symptom **3030** for high grass. Use the HIGH or FAST control setting.

Did the mower perform satisfactorily?

YES—Fault has been isolated: Incorrect carburetor adjustment.

NO—Fault may exist in other parts of the engine's symptoms. Proceed to Fault Symptom **3420.**

ENGINE RUNNING PROBLEM LACK OF POWER, EXHAUST SYSTEM FAULTY

FAULT SYMPTOM 3420

Possible Causes:

- Wrong muffler on engine ——————————— **See 3421**

- Clogged muffler ————————————————— **See 3422**

- Clogged exhaust ports on 2-cycle engine ——— **See 3423**

Initial Conditions Check List:

Perform all the pre-start checks and conditions given in the Check List of Fault Symptoms **1000** and **2000**.

3421

Wrong muffler on engine

Step 1—The wrong size muffler on an engine will cause the engine to perform improperly. In particular, if a smaller than specified muffler has been used the engine will lack power.

If the muffler has been replaced, check with the manufacturer's specification to insure an exact replacement was made. If the wrong size muffler has been used, replace the muffler with the specified size as follows:

a. Place the control to OFF or STOP.

b. Make sure the engine and muffler are cool to the touch, then carefully remove the muffler from the engine.

c. Usually the muffler can be unscrewed by hand. If not, use a wrench at the end closest to the engine. Be careful not to damage the muffler. If the muffler does not move, apply a few drops of penetrating oil or liquid wrench (which is a very light oil used on rusty parts to loosen them) to the threaded part or parts going into the engine block. Wait a few minutes, for the oil to penetrate and then unscrew the muffler.

d. Install the new muffler. Be sure it is tight.

e. Start up and run engine until engine warms up. Repeat the power test outlined in Fault Symptom **3030** for high grass. Use the HIGH or FAST setting.

Did the mower perform satisfactorily?

YES—Fault has been isolated and corrected: Wrong muffler on engine.

NO—Proceed to Fault Symptom **3422**.

3422

Clogged muffler

Step 1—An accumulation of carbon inside the engine muffler will reduce the power output of the mower engine. Refer to Fault Symptom **3421** for the procedure for removal of a muffler suspected of being clogged.

After removal of the muffler, examine it visually. If the openings appear to be clogged with carbon or exhaust deposits, discard the muffler and replace it with a new one. Be sure you have the identical replacement muffler. Upon completion, start up and run the engine. Permit warm up, then run the power test per Fault Symptom **3030**. Use a HIGH or FAST control setting.

Did the mower perform satisfactorily?

YES—Fault has been isolated and corrected: Clogged muffler.

NO—For 2-cycle engines, proceed to Fault Symptom **3423**. For 4-cycle engines proceed to other engine system checks in Fault Symptom **3430**.

3423

Clogged exhaust ports on 2-cycle engine

Step 1—When carbon and exhaust deposits build up on the exhaust ports of a 2-cycle engine the efficiency, and hence the power, will begin to drop. Refer to Fault Symptom **2641** or **2642** as appropriate (for horizontal or vertical mounted 2-cycle engine) for the procedure for clean out of carbon on the engine's exhaust ports.

When the clean out is completed and the engine ready to run, start up and run the engine.

Permit engine warm up, then run the power test per Fault Symptom **3030**. Use a HIGH or FAST control setting.

Did the mower perform satisfactorily?

YES—Fault has been isolated: Clogged exhaust ports.

NO—Proceed to Fault Symptom **3430** for other engine system checks.

ENGINE RUNNING PROBLEM
LACK OF POWER,
IGNITION SYSTEM FAULTY

FAULT SYMPTOM   3430

Possible Causes:

- Loose connection—ignition cable to plug ————— **See 3431**

- Defective spark plug ————————————— **See 3432**

- Weak or poor quality spark ————————— **See 3433**

Initial Conditions Check List:

Perform all the pre-start checks and conditions given in the Check List of Fault Symptoms **1000** and **2000**.

3431

Loose connection—ignition cable to plug

Shut down the engine, position THROTTLE to OFF. Examine the connection made by the ignition cable at the spark plug terminal. The connection should be tight. If the end of the ignition cable cap wobbles around or is very loose on the plug terminal, this can result in engine power loss. Push the end cap all the way down when attaching the cable. If the cap lead screws onto the terminal, attach securely. If the cap lead has an insulating boot over it and the inside cap fits the plug terminal very loosely, carefully crimp down on the outside of the boot using a pair of smooth jaw pliers. Do not crimp too hard such that the cap lead no longer fits over the plug terminal. Crimp just enough so that the cap fits tightly onto the terminal. Attach the ignition cable and start up engine.

Perform the engine power test per Fault Symptom **3030.** Use a HIGH or FAST control setting.

Did the mower perform satisfactorily?

YES—Fault has been isolated: Loose ignition cable connection on spark plug.

NO—Proceed to Fault Symptom **3432.**

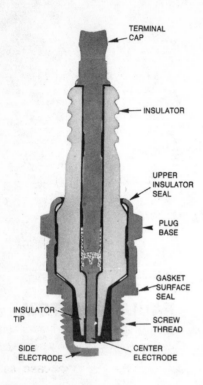

TERMINAL CAP

INSULATOR

UPPER INSULATOR SEAL

PLUG BASE

GASKET SURFACE SEAL

INSULATOR TIP

SCREW THREAD

SIDE ELECTRODE

CENTER ELECTRODE

FIGURE 116 Cutaway View of Typical Spark Plug

3432

Defective spark plug

Step 1—Lack of engine power can be due to a defective or dirty spark plug. Remove the spark plug per Service Procedure **4004**. Refer to Fig. 116 for a cut-away view of a typical spark plug for electrode location, etc.

a. Examine the spark plug by referring to Fault Symptoms **2220** and **2230**.

b. After cleaning, gapping, or replacing the plug with a new plug, reinstall the plug and connect the ignition cable.

c. Start up and run the engine.

d. Permit engine warm-up. Then perform the engine power test per Fault Symptom **3030**. Use a HIGH or FAST control setting.

Did the mower perform satisfactorily?

YES—Fault has been isolated: Defective spark plug.

NO—Proceed to Fault Symptom **3433**.

3433

Weak or poor quality spark

Step 1—When the ignition system is providing a weak spark, or a poor spark due to improper voltage generation, the engine will not be able to generate maximum power.

Refer to Fault Symptom **2240** and perform the test of **2240** to determine the quality of the spark and capability of the basic magneto-electrical system to generate the proper spark.

Did the test result in a spark with the special spark test plug?

YES—The fault is not verified. Proceed to Fault Symptom **3440** for other engine system checks.

NO—Fault has been isolated: Weak or poor quality spark will require overhaul of the ignition system. Refer to Service Procedure **4030**.

ENGINE RUNNING PROBLEM
LACK OF POWER, VALVES
OR SEALS FAULTY

FAULT SYMPTOM   3440

Possible Causes:

- Poor compression, leaky valves/weak springs,

  4-cycle engine ——————————————— **See 3441**

- Poor compression, leaky crankcase seal,

  2-cycle engine ——————————————— **See 3442**

- Poor compression, leaky ignition seal ————— **See 3443**

Initial Conditions Check List:

Perform all the pre-start checks and conditions given in the Check List of Fault Symptoms **1000** and **2000**.

3441

Leaky valves/weak springs, 4-cycle engine

Step 1—On 4-cycle engines if the intake/exhaust valves leak, or if the valve springs become weak, a loss in engine power will occur. The most conclusive test of the proper performance of these valves is to run an engine compression test. Refer to Fault Symptom **2550**. Note specifically the determinations of **2550**, Step 2, with regard to the condition of the intake/exhaust valves.

Do the compression gage readings show open or sticky valves?

YES—Fault has been isolated and verified: Leaky valves or weak valve springs. Refer to Service Procedure **4100**.

NO—Proceed to Fault Symptom **3443**.

3442

Leaky crankcase seal, 2-cycle engine

Step 1—A leaky crankcase seal on a 2-cycle engine can result in a loss of engine power. Refer to Fault Symptom **2610** or **2620** (rotary or reel mower engines) and perform the tests for a leaky seal.

Did the test indicate the crankcase seal was leaking?

YES—Fault has been isolated: Leaky crankcase seal. Refer to Service Procedure **4100**.

NO—Proceed to Fault Symptom **3443**.

3443

Leaky ignition seal

Step 1—A leaky ignition seal for either a 2- or 4-cycle engine can eventually result in loss of engine power due to contamination of the ignition system. Specifically, when the ignition breaker points begin to accumulate any fuel-oil mixture, or oil deposits, their electrical properties will deteriorate.

Refer to Fault Symptom **2630** for a procedure for examining the condition of the ignition seal.

Did the examination reveal oil or raw fuel around the ignition seal?

YES—Fault has been isolated: Leaky ignition seal. Refer to Service Procedure **4035**.

NO—The fault is not verified. Proceed to Fault Symptom **3450** for other engine system checks.

ENGINE RUNNING PROBLEM
LACK OF POWER, SPEED
GOVERNOR FAULTY

FAULT SYMPTOM   3450

Possible Causes:

- Governor linkage jam ──────────────── **See 3451**
- Governor requires adjustment ──────────── **See 3452**

Initial Conditions Check List:

Perform all the pre-start checks and conditions given in the Check List of Fault Symptom **2000**.

3450

General types of engine governors

Step 1—When the engine power performance test of **3030** shows the engine lacks power, and the engine exhibits difficulty in holding power settings with the control lever positioned at HIGH or FAST, the governor may be the problem. Some typical governor linkage arrangements are shown in Figs. 117, 118 and 119. Note that there is a specific relationship between the throttle linkage and the governor linkage.

Step 2—The governor on an engine is used to control engine speed from no load to full load conditions. In addition the governor prevents the engine from overspeeding or from developing a runaway speed.

Runaway speed can literally tear the engine apart. There are 2 basic types of governors used for mower engines. One is an air vane governor and the other a mechanical flyweight or flyball. The air vane governor operates by means of an air blast created by the finned flywheel moving the air vane. See Fig. 118. Movement of the vane is restrained and controlled by a light spring and the air blast from the flywheel. The vane is connected to the throttle linkage and when a load is applied to the engine the flywheel slows down causing less of an air blast. At this point the light spring moves both the vane and the linkage in proportion to the engine speed slowdown. As the speed begins to slow down, the vane and throttle linkage will open the throttle equal to the amount of speed lost, and the engine then will run faster. In actual operation it is difficult to even notice these constantly changing conditions, they occur smoothly and quickly.

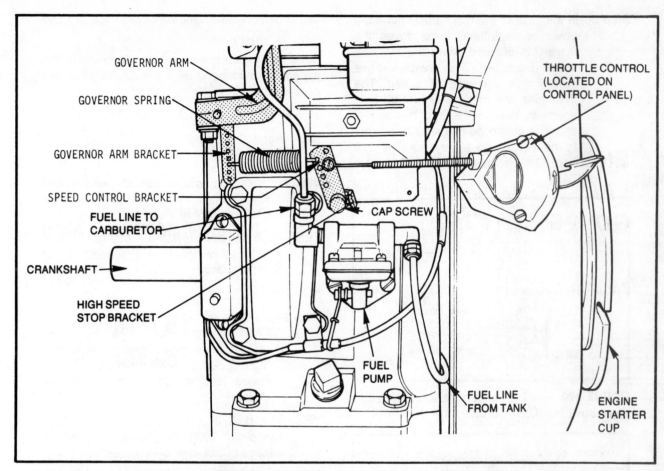

GOVERNOR ARM

GOVERNOR SPRING

GOVERNOR ARM BRACKET

SPEED CONTROL BRACKET

FUEL LINE TO
CARBURETOR

CRANKSHAFT

HIGH SPEED
STOP BRACKET

THROTTLE CONTROL
(LOCATED ON
CONTROL PANEL)

CAP SCREW

FUEL
PUMP

FUEL LINE
FROM TANK

ENGINE
STARTER
CUP

FIGURE 117   Typical Variable Speed Governor Arrangement

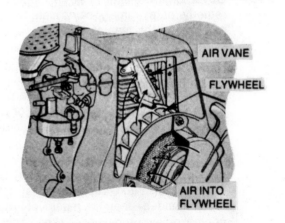

AIR VANE

FLYWHEEL

AIR INTO
FLYWHEEL

FIGURE 118   Typical Air Vane Type Governor

GOVERNOR
SECURING
SCREW

GOVERNOR
MOUNTING
SCREW

FIGURE 119   Governor Positive Control Type Linkage and Adjusting Screw

Step 3—In the case of the mechanical flyweight governor centrifugal force drives the flyweights outward when the flywheel speed increases. The weights move inward with decrease of speed. The flyweights are attached—through linkages—to the throttle and the engine speed is controlled in a similiar manner to the air vane governor. See Figs. 120 and 121. Proceed to Fault Symptom **3451**.

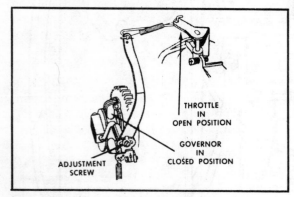

FIGURE 120   Typical Mechanical Flyweight Type Governor, Open Throttle

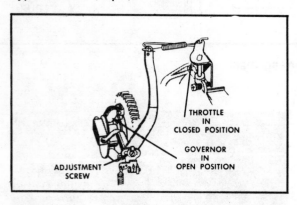

FIGURE 121   Mechanical Flyweight Type Governor, Closed Throttle

3451

Governor linkage jam

Step 1—When the engine power exhibits difficulty in speed control the engine speed governor may be jammed. Proceed as follows:

a.   Start up and run the engine.

b.   Position the engine control to IDLE or LOW SPEED.

Does the engine speed correspond to idle or low speed, as evidenced by a low sound which the engine makes?

YES—Proceed to Step 3.

NO—Proceed to Step 2.

Step 2—Does the engine run at high or fast speed even though the control is at IDLE or LOW SPEED?

YES—Fault has been isolated: Governor linkage is jammed or defective. Proceed to Fault Symptom **3452.**

NO—Proceed to Step 3.

Step 3—Position the speed control to HIGH or FAST.
      Does the engine speed correspond to high speed as evidenced by the high speed loud sound which the engine makes?

YES—The fault is not verified. Refer to Fault Symptom **3000** for listing of possible causes of lack of engine power.

NO—Fault has been isolated: Governor linkage is jammed or defective. Proceed to Step 4.

Step 4—Governor linkage jams can be simple to remedy or extremely complicated depending on the type of governor. Refer to the figures and discussion of the 2 general types of governors given in Fault Symptom **3450** and then determine whether to proceed to dismantle the engine to unjam the governor or to take the mower to an Authorized Factory Service Dealer.

      In cases where the reader is not completely familiar with the use of engine tools and repairs, it is not recommended that a repair of the engine governor be undertaken. Additionally, governor jams usually involve the removal and replacement of the calibrated governor spring. If care is not exercised this spring can be damaged with the result that the mower will be very troublesome in operation.

3452

Governor requires adjustment

Step 1—Adjustment of the governor for different engine designs varies in both convenience and complexity. One convenient and simple design is on the Jacobsen 321 engine shown in Fig. 119. A different governor is discussed in Step 3. In the Jacobsen governor the linkage is made up of arms, linkages, spring and adjustment screw. Illustration of the governor adjusting mechanism is shown in Fig. 119. In this view the carburetor has been removed. Positive governor control can be obtained on this engine.

The adjusting screw "c" of Fig. 119 allows increase or decrease of engine speed of about 400 RPM. Before using this governor adjustment, the engine speed must be determined accurately using a Tachometer. Proceed as follows:

a. Be certain that the THROTTLE control cable has been adjusted properly. Refer to Fault Symptom **3402**. The control panel position markings should correspond to the proper carburetor lever positions.

b. Position the THROTTLE control lever to OFF.

c. Obtain and connect up a simple Tachometer as shown in the accompanying Fig. 122.

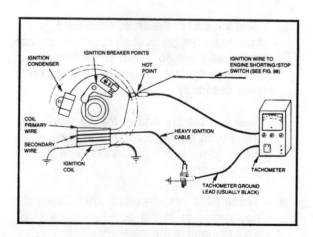

FIGURE 122 Tachometer Hook-up Connections

d. On most engines the hot lead wire is the smaller diameter of the two wires which come out from under the magneto flywheel. The heavy wire is the high tension ignition cable. This wire goes to the spark plug. The hot lead wire usually goes to the starter switch or to the control lever, or a point near the carburetor where movement of the control cable to the STOP position shorts the hot lead wire (voltage) to ground. The black Tachometer wire is ground and can be connected to any convenient ground point. Use of the spark-plug base is a preferred point.

e. Refer to Service Procedure **4050** for a listing of typical manufacturer's maximum, and idle RPM. A top speed of 3600 RPM will be selected as an illustrative example here.

f. Start up engine, let engine warm up at IDLE control lever position for about 5 minutes, meanwhile read the RPM on the Tachometer.

g. Position the control lever to FAST or HIGH, read the RPM, move the lever slightly to obtain the highest RPM.

h. If the *maximum* RPM possible is (assumed) 3600 RPM, turn the governor adjusting screw until the RPM reads the desired 3600 RPM (which is the recommended RPM for top speed).

Step 2—Shut down the engine, position control lever to OFF. Disconnect and remove the Tachometer. Start up the engine and perform the engine power test per Fault Symptom **3030**.

Does the mower perform satisfactorily with the engine providing the necessary power to cut high grass?

YES—Fault isolated and corrected: Governor setting required adjustment.

NO—The possibility exists for multiple faults occurring simultaneously. Review Fault Symptom **3400** and attempt to correlate engine problem symptoms with specific faults.

Step 3—Another simple governor adjustment design which is different from that given in Step 1 is one incorporated on Kohler single cylinder lawn mower engines. A fixed speed governor is shown in Fig. 123. This governor is a mechanical centrifugal flyweight type mounted within the crankcase. A gear on the crankshaft drives the mechanism. This governor is factory adjusted but if the arm or linkage works loose, it will require adjustment. Proceed as follows for initial adjustments and final speed adjustments:

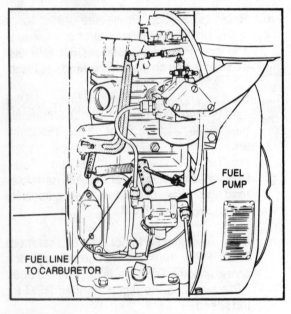

FIGURE 123   Typical Fixed Speed Type Governor

NOTE:

Kohler engine models K241, K301, K321 use a fixed speed type governor. An initial adjustment on the governor is required (refer to Fig. 123).

a. Position the THROTTLE control lever to OFF. If the engine is hot, wait until it cools.

b. Hold the governor cross shaft, as shown in Fig. 124 and loosen the arm hex nut which holds the governor arm to the governor cross shaft.

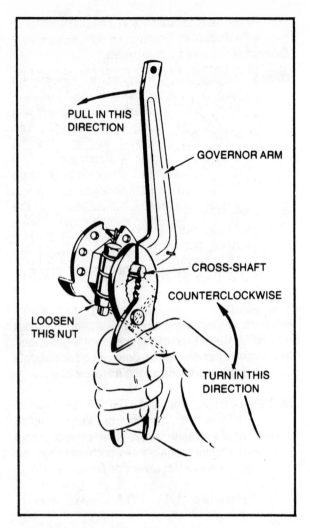

FIGURE 124   Initial Adjustment for Kohler Fixed Speed Governor

c. Hold the end of the cross shaft per Fig. 124 and rotate the cross shaft counterclockwise as far as it will go. A tab on the cross shaft will stop against the rod on the governor gear assembly.

d. Now pull on the end of the governor arm, away from the carburetor. Use a moderate force, do not pull with a heavy force so as to bend the arm.

e. Tighten the arm hex nut which holds the governor arm to the governor cross shaft. This completes the initial adjustment.
    After completing the initial adjustment a final speed adjustment is required.

f. Refer back to Step 1 for use and hook-up of a Tachometer. Connect up the Tachometer.

g. Start engine, permit 5 minute warm-up and position the THROTTLE lever control to FAST or HIGH. Read RPM. The Kohler K241, K301, and K321 engines have a *maximum* allowable speed of 3600 RPM. Move the THROTTLE slightly back and forth until a maximum RPM reading is obtained.

h. If the RPM is found to be *low*, first loosen the lock nut, then tighten the speed adjusting nut shown on Fig. 123. Tighten nut for a maximum 3600 RPM. If the RPM is higher than 3600 loosen this adjusting nut until a reading of 3600 is obtained.

i. When the proper RPM is obtained tighten up on the lock nut and recheck the RPM. This process may have to be repeated a few times until the exact RPM is obtained.

NOTE:

This procedure was applicable to Kohler engines with *fixed speed governors*. For engines with *variable speed governors* as shown in Fig. 117 the initial adjustment applies. In addition a final speed adjustment requires use of a Tachometer. However the final speed adjustment is similar and is accomplished by simply loosening the capscrew shown in Fig. 117, rotating the high speed stop bracket until the reading of 3600 RPM is obtained, and tightening the capscrew. Disconnect the Tachometer and Proceed to Step 4.

Step 4—Operate the mower and perform the engine power test per Fault Symptom **3030**.

Does the mower perform satisfactorily with the engine providing the necessary power to cut high grass?

YES—Fault isolated and corrected: Governor setting required adjustment.

NO—The possibility exists for multiple faults occuring simultaneously. Review Fault Symptom **3400** and attempt to correlate engine problem symptoms with specific faults.

ENGINE RUNNING PROBLEM
ENGINE RUNS AT MID TO HIGH
SPEEDS—STALLS AT IDLE SPEED FAULT SYMPTOM 3500

Possible Causes:

- Restricted or obstructed air supply ——————— **See 3510**

- Improperly adjusted or dirty carburetor

  idle adjustment ———————————————— **See 3520**

- Carburetor restricting fuel supply ————— **See 3530**

- Faulty spark plug ————————————— **See 3540**

- 2-cycle engine—improperly performing

  reed valve —————————————————— **See 3550**

- Incorrectly set speed governor ————— **See 3560**

Initial conditions check list:

Perform all the pre-start checks given in the Check Lists of Fault Symptoms **1000** and **2000**.

3510

Restricted or obstructed air supply

Step 1—When the carburetor air supply is partially shut off due to a clogged air cleaner or obstruction in the air intake the engine will not idle.

Shut down the engine, and position the THROTTLE to OFF. Proceed as follows:

a. Remove the air cleaner and filter element, and if the element can be cleaned, service as shown in Figs. 125 to 129.

b. Follow the procedure given in the figures for adding a light film of oil to the filter element.

c. Clean out the filter container.

d. Before replacing the filter container, look into the carburetor. Look for dirt and debris such as grass clippings and the like. Clean out any dirt found. The carburetor must be completely clean.

e. Reinstall the element (or replace as required) and replace the air cleaner container on the engine (if it was necessary to remove the container).

Start up the engine, permit warm up at a midspeed position, then position control lever to IDLE or SLOW.

Does the mower engine idle satisfactorily?

YES—Fault has been isolated: Carburetor air supply partially clogged.

NO—Proceed to Fault Symptom **3520.**

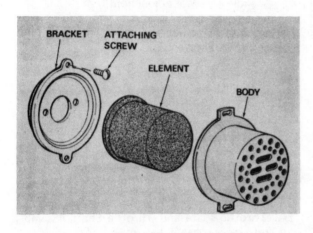

FIGURE 125   Removal of Air Filter

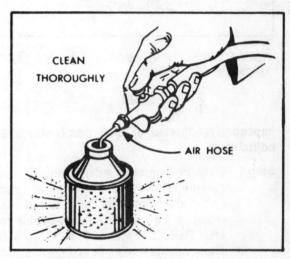

FIGURE 127   Dry Paper Type Filter

FIGURE 126   Cleaning Metallic Filter

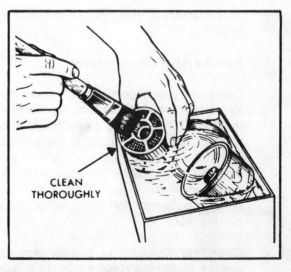

FIGURE 128   Oil Bath Type Air Cleaner

FIGURE 129 Polyurethene Element Type
Air Cleaner

3520

**Improperly adjusted or dirty carburetor idle
adjustment**

Step 1—Dirty or improper adjustment of the
carburetor idle fuel flow screw (also
called idle speed regulating screw) can
result in the engine not being able to
idle. Refer to Figs. 130 to 137 which
show various makes of carburetors and
identify the location of the carburetor
idle adjustment screw. Proceed as
follows:

a.   Shut down the engine, position THROTTLE
to OFF.

b.   If engine is hot, permit it to cool.

c.   Using only your fingers, carefully screw the
idle screw out of the carburetor. Examine
the screw. There should be no evidence of
wear or rings/indentations at the pointed
end of the screw. If there are, and the tip is
indented or appears worn, replacement is
necessary. (Refer to exploded line drawings
of your carburetor for exact replacement
part number.)

d.   Clean the screw if it appears dirty or
gummy. Carburetor cleaner should be used
for cleaning the screw.

e.   After cleaning the screw, carefully screw it
back into the carburetor with your fingers.
Turn it in until it is fully seated and finger
tight. Do not force the screw! It can be
damaged easily.

f.   Now back the screw out one full turn.

Start up the engine, permit warm up at a
midspeed position, then position control lever to
IDLE or SLOW.

Does the mower engine idle satisfactorily?

YES—Fault has been isolated: Improper/dirty
idle screw or adjustment.

NO—Proceed to Fault Symptom **3530.**

A. Throttle Plate and Throttle Shaft Assembly
B. Idle Speed Regulating Screw
C. Idle Fuel Regulating Screw
D. Choke Plate and Choke Shaft Assembly
E. Atmospheric Vent Hole
F. Float Bowl Housing Drain Valve
G. Float Bowl Housing Retainer Screw
H. High-speed Adjusting Needle
J. Float Bowl Housing
K. Idle Chamber

FIGURE 130 Side Draft, Float Bowl Type
Carburetor

3530

Carburetor restricting fuel supply

Step 1—When the lawn mower engine will not run
at idle speed, it may be due to lack of
fuel being supplied by the carburetor.
This condition can occur when a carburetor with a float and needle valve
arrangement (see Figs. 138 and 139 for
typical float and needle valve carbure-

tors) is not properly adjusted. The float level adjustment may not permit a sufficient or proper fuel flow. As a result the engine will not idle.

There is no convenient external adjustment which can be performed for the float level. The carburetor must be removed, float bowl removed or disassembled, and the float tested for proper adjustment. Details on this operation are given in Service Procedure **4010**.

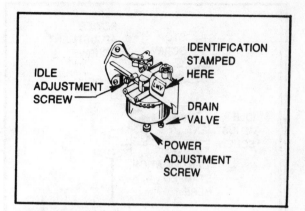

FIGURE 133   Type LMV Carburetor

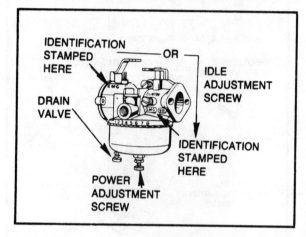

FIGURE 131   Type LMG Carburetor

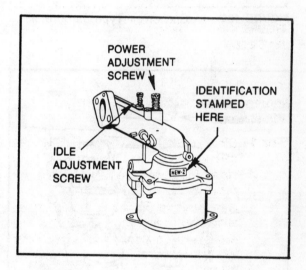

FIGURE 134   Type HEW Carburetor

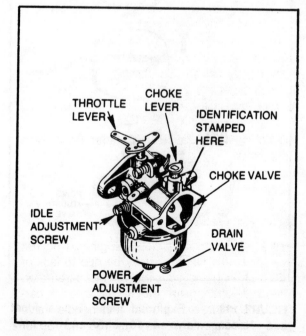

FIGURE 132   Type LMB Carburetor

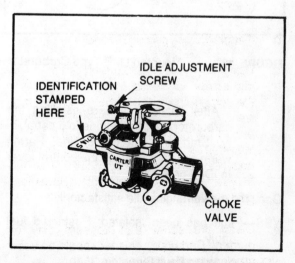

FIGURE 135   Type UT Carburetor

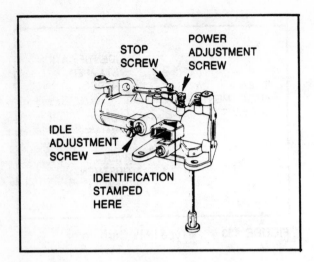

FIGURE 136   Suction Lift Type Carburetor

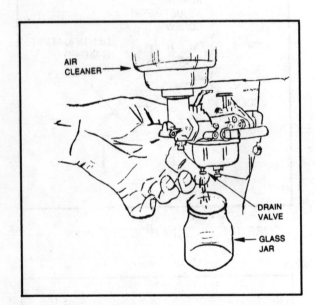

FIGURE 137   Mounted LMB Type Carburetor

After servicing and re-installing the carburetor, test the engine idle capability. Start and run the engine. Permit warm-up, then position the control lever to IDLE or SLOW.

Does the mower engine idle satisfactorily?

YES—Fault has been isolated: Restricted fuel supply.

NO—Proceed to Fault Symptom **3540.**

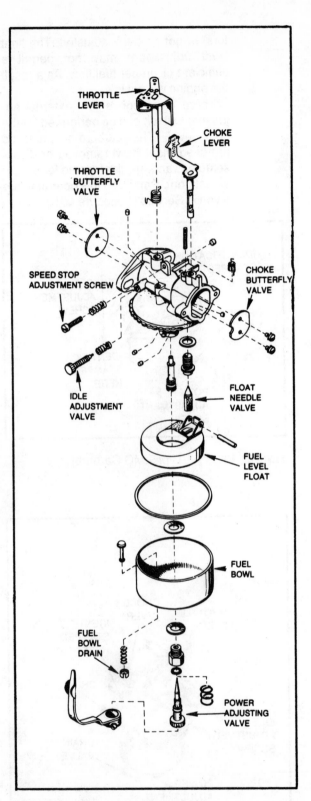

FIGURE 138   Exploded View, Typical Float and Needle Valve Type LMG, LMB, LMV Carburetor

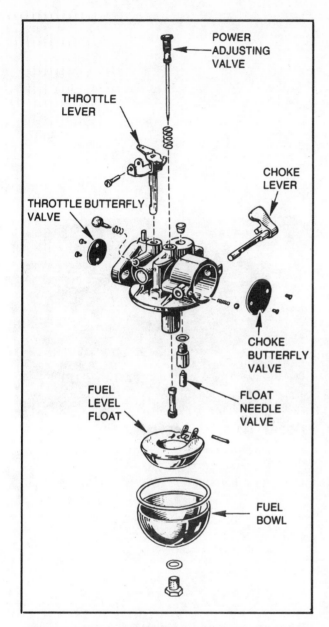

FIGURE 139 Exploded View, Typical Float and Needle Valve Carter Carburetor

3540

Faulty spark plug

Step 1—The component of the ignition system that will most likely prevent the engine from idling is the spark plug. If the spark-plug gap is set too close and not in accordance with the manufacturer's specifications, the engine will have difficulty in idling. Proceed as follows:

a. Shut down engine, position THROTTLE lever to OFF. Allow engine to cool.

b. Remove spark plug, measure gap and regap the plug as shown in Service Procedure **4004**. Note that the gap will vary with the manufacturer, and whether the engine is a 2- or 4-cycle engine. Consult your engine manufacturer's specification for proper gap setting. If this information is not available, a general setting of .028 inches may be used.

c. If the plug is dirty, be sure to clean per Service Procedure **4004** before regapping.

d. Reinstall the plug and connect ignition cable.

Start and run the engine, permit warm-up at a midspeed position, then position lever to IDLE or SLOW.

Does the mower engine idle satisfactorily?

YES—Fault has been isolated: Faulty spark-plug gap setting.

NO—Proceed to Fault Symptom **3550** for 2-cycle engine or to **3560** for either 2- or 4-cycle engines.

3550

2-cycle engine—improperly performing reed valve

Step 1—When the mower engine runs but stalls when IDLE speed is selected, it may be suspected that there is a leaky reed valve. This valve admits, or restricts, the fuel air mixture flow into the crankcase part of the engine as shown in the sketch of Fig. 140. The reed valve opens and closes in accordance with the movement of the engine piston as shown in Figs. 141, 142 and 143. It is evident from these sketches that if the reed valve does not work properly, the engine will not perform well.

To test the reed valve operation, proceed as follows:

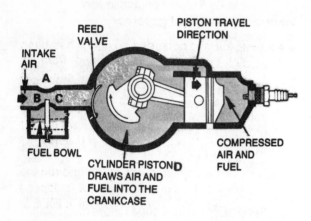

FIGURE 140   Typical Reed Valve Operation, 2-Cycle Engine

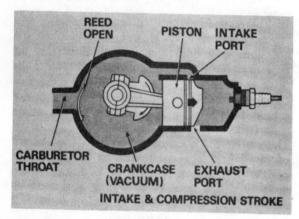

FIGURE 141   Reed Valve Opens, Admits Fuel/Air

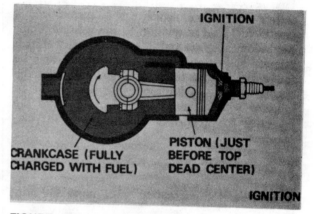

FIGURE 143   Ignition Phase

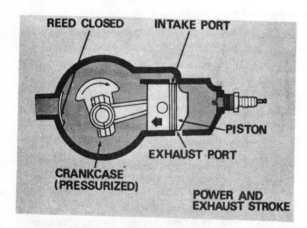

FIGURE 142   Reed Valve Closes

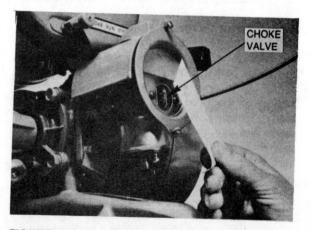

FIGURE 144   Test For Leaky Reed Valve

a. Place controls or THROTTLE at OFF or STOP.

b. Remove the carburetor air cleaner assembly.

c. Start the engine, and with the engine running, hold a 1½ to 2 inch wide piece of white clean paper about 1 inch from the intake of the carburetor as shown in Fig. 144.

d. Examine the paper. If the paper has been spotted by the fuel mixture, then the reed valve is leaking and not seating properly.

e. Shut down engine.

Was there evidence of the reed valve leaking?

YES—Fault has been isolated: Leaky reed valve. Refer to Service Procedure **4012**.

NO—Shut down engine and proceed to Fault Symptom **3560**.

3560

Incorrectly set speed governor

Step 1—If the engine speed governor has been set too fast or has slipped from its proper setting, then this fault can keep the engine from idling properly. Refer to Fault Symptom **3452** for discussion on proper adjustment of typical lawn mower engine speed governors and adjust as applicable.

After adjustment, start up and run the engine. Permit warm-up at a midspeed position, then position control to IDLE or SLOW.

Does the mower engine idle satisfactorily?

YES—Fault has been isolated: Incorrectly set speed governor.

NO—Review Fault Symptom **3500** for other possible causes of inability of engine to idle properly.

ENGINE RUNNING PROBLEM
ENGINE IDLES—BUT
IDLES POORLY

FAULT SYMPTOM   3570

Possible Causes:

- Incorrect idle adjustment ——————————— **See 3571**
- Air leak into engine cylinder ——————————— **See 3572**
- Carburetor restricting fuel supply ——————— **See 3573**

Initial Conditions Check List:

Perform all the pre-start checks given in the Check List of Fault Symptoms **1000** and **2000**.

3571

Incorrect idle adjustment

Step 1—If the mower engine is capable of idling, but idles poorly, having great difficulty idling, then the idle adjustment is not correct. Refer to Fault Symptom **3520** and follow the procedure given for proper adjustment, and cleaning, of the carburetor idle screw.

Upon completion of this procedure start up and run the engine. Run at midspeed until warmed up, then position control at IDLE.

Does the mower engine idle satisfactorily?

YES—Fault has been isolated: Incorrect Symptoms adjustment.

NO—Proceed to Fault Symptom **3572.**

3572

Air leak into engine cylinder

Step 1—An engine which has developed an air leak between the carburetor and cylinder, or intake manifold, will change the air-to-fuel ratio. This will result in a very lean fuel mixture and contribute to poor idling. For 2-cycle engines this condition is particularly bad. Proceed as follows:

a. Shut down the engine; permit to cool, if hot.

b. Clean off the engine carefully, removing all grime and dirt, particularly around the carburetor.

c. Examine the surfaces which are joined together. Note particularly the joint which is on the carburetor and which attaches to the cylinder.

d. Start up the engine, and run at high speed.

e. Look for any evidence of fuel bubbling out around the carburetor attaching joint, and at any other point on the carburetor.

Is there any fuel bubbling out around the carburetor attaching joint?

YES—Fault has been isolated: Air leak into engine cylinder. Proceed to Step 2.

NO—Proceed to Fault Symptom **3573**.

Step 2—Shut down the engine and tighten up on the bolts or screws which hold the carburetor to the cylinder.

Start up and run the engine and observe if fuel still is bubbling out of the joint.

Does fuel still bubble out?

YES—Fault has been isolated. Refer to Service Procedure **4011**.

NO—Proceed to Step 3.

Step 3—Start up and run the engine. Position the control lever at IDLE speed.

Does the engine idle satisfactorily?

YES—Fault isolated and verified: Air leak into cylinder.

NO—Proceed to Fault Symptom **3573**.

3573

Carburetor restricting fuel supply

Step 1—If the lawn mower engine idles poorly, it may be due to an insufficient fuel supply to the carburetor.

This condition can occur when a carburetor with a float and needle valve arrangement (see Figs. 138 and 139 for typical float and needle valve carburetor) is not properly adjusted. The float level adjustment may not permit a sufficient or proper fuel flow. As a result, the engine will not idle.

Refer to Fault Symptom **3530** for discussion on proper adjustment of the float valve.

After adjustment, start up and run the engine. Permit warm-up. Then position control to IDLE or SLOW.

Does the engine idle satisfactorily?

YES—Fault has been isolated: Carburetor restricting fuel supply.

NO—Refer to Fault Symptom **3500** for other symptoms on poor idling.

ENGINE RUNNING PROBLEM
ENGINE STARTS—BUT
RUNS ONLY AT SLOW SPEED   FAULT SYMPTOM   3580

Initial Conditions Check List:

Perform all the pre-start checks of the Check Lists of Fault Symptoms **1000** and **2000**.

Step 1—This problem occurs with lawn mower engines that have been operating for several seasons without proper maintenance. The mower starts up but runs very slowly no matter where the control lever is positioned. A slow putt-putt sound is made by the engine and there is obvious lack of power. This problem will most likely occur when first starting up the mower for the cutting season and is caused by the choke valve being stuck almost closed. Proceed as follows:

a. Shut down engine, position THROTTLE to STOP.

b. If engine is hot, allow to cool.

c. Remove air cleaner and container so that the choke valve butterfly can be seen. The valve should be in the fully closed position.

d. Use a spray can of carburetor control linkage cleaner and spray the choke valve, particularly the area around the butterfly valve and the valve shaft. A clean cloth can be used to remove any dissolved gum or varnish. Be careful not to leave any lint or threads in the carburetor.

Step 2—Spray a second time: Work the control lever back and forth making sure the choke valve opens and closes freely in response to the control lever. If this procedure fails to free the choke valve refer to Fault Symptom **2110**.

Step 3—Permit the carburetor to dry, then replace air cleaner and container and start up engine.

Does the mower engine operate at all control lever positions, and with the engine evidencing proper response to the control?

YES—Fault has been isolated: Choke butterfly valve stuck almost closed.

NO—Proceed to Step 4.

Step 4—Was there any improvement in the engine being able to run at more than just idle speed?

YES—The fault has been isolated, but includes other faults beyond a stuck choke valve. Refer to Fault Symptom **3500** for other symptoms on poor idling.

NO—The carburetor must be removed from the engine and thoroughly cleaned. Refer to Service Procedure **4010**.

ENGINE RUNNING PROBLEM
ENGINE RUNS UNEVENLY—
SURGES

FAULT SYMPTOM   3600

Possible Causes:

- Water or dirt in fuel ————————————————— See 3610
- Wrong fuel (mixture) ——————————————————— See 3620
- Plugged fuel tank vent ——————————————————— See 3630
- Sticky throttle valve or sticky choke valve ————— See 3640
- Sticky governor, or faulty governor ——————— See 3650
- Engine is loose ———————————————————————— See 3660
- Damaged or bent crankshaft ——————————————— See 3670

Initial Conditions Check List:

Perform all the pre-start checks given in the Check Lists of Fault Symptoms **1000** and **2000**.

3610

Water or dirt in fuel

Step 1—If water or dirt has accumulated in the fuel supply system, the result can be that the engine slows, speeds up, slows and repeats this, and generally gives poor performance.

Refer to Fault Symptom **3110** for a procedure for checking if water or dirt has gotten into the fuel tank or fuel line. After completing the test, is there any evidence of water or dirt in the fuel?

YES—Fault has been isolated. Proceed to Step 2.

NO—Proceed to Fault Symptom **3620**.

Step 2—After clean out of the fuel system per Symptom **3110** refill with proper fresh fuel.
Start up and run the engine.

Does the engine run satisfactorily without surging?

YES—Fault has been verified and corrected.

NO—Proceed to Fault Symptom **3620**.

3620

Wrong fuel (mixture)

Step 1—For 2-cycle engines it is necessary that the proper mixture of gasoline and oil is being used per the manufacturer's specification. See Table I for typical manufacturer's mixtures. If there is any doubt of the proper mixture, drain the fuel tank and refill with proper mixture. Start up and run engine.

Does the engine run satisfactorily without surging?

YES—Fault has been isolated: Improper fuel mixture.

NO—Proceed to Fault Symptom **3630.**

3630

Plugged fuel tank vent

Step 2—When the fuel tank vent in the cap is becoming plugged the engine will be starved for fuel. Surging can then occur. Refer to Fault Symptom **3120** for a procedure for checking if the tank cap vent is plugged. After completion of Symptom **3120,** test and clean out or replace the tank cap, start up and run the engine.

Does the engine perform satisfactorily without surging?

YES—Fault has been isolated: Plugged fuel cap vent.

NO—Proceed to Fault Symptom **3640.**

3640

Sticky throttle valve or sticky choke valve

Step 1—A problem which can occur when the lawn mower is not maintained on a regularly scheduled basis is the throttle or choke butterfly valve sticks or the linkage shaft binds. When these parts of the carburetor do not move freely and easily, engine surge can develop. Proceed as follows:

a. Shut down engine, position throttle to STOP.

b. If engine is hot, allow to cool.

c. Remove air cleaner and container so that the choke valve butterfly can be seen. It should be in the fully closed position.

d. Use a spray can of carburetor control linkage cleaner and spray the choke valve, particularly the area around the butterfly valve and the valve shaft. A clean cloth can be used to remove any dissolved gum or varnish. Be careful not to leave any lint or threads in the carburetor.

e. Spray a second time. Work the control lever back and forth while making sure the choke valve opens and closes freely in response to the control lever.

f. Position the throttle to RUN or FAST.

g. Spray again. Direct the spray into the carburetor throat towards the throttle valve inside.

h. Permit the carburetor to dry.

i. Replace air cleaner and container and start up engine. Permit warm-up, then work the control lever back and forth.

Does the engine perform satisfactorily without surging?

YES—Fault has been isolated: Sticky throttle or choke valve.

NO—Proceed to Fault Symptom **3650.**

3650

Sticky governor, or faulty governor

Step 1—An improperly operating governor linkage or arm will cause the engine to surge. Proceed as follows:

a. Shut down engine, position throttle to STOP. If engine is hot, allow to cool down completely.

b. Remove the air cleaner container, and any cover or shroud so that the carburetor throttle linkage to the governor is accessible. Refer to Figs. 145, 146, 147 and 148 for some views of typical governor linkages.

c. Use a spray can of carburetor control linkage cleaner and spray the governor linkages and throttle linkages. A clean cloth can be used to wipe dry. Be careful not to leave lint or bits of cloth on the linkages.

d. Examine the governor arm and note if the arm has been damaged or bent. If the arm has been bent, attempt to straighten. If in doubt as to how it should appear when straightened, discard the arm and install a new factory supplied arm identical to the original.

e. Work the control lever back and forth and see if all the linkages follow for high and low speed settings. There should be no sticking or binding on any linkage or part.

f. Replace the air cleaner, and any shroud or cover which has been removed.

Start up and run the engine. Does the engine run satisfactorily, without surging?

YES—Fault has been isolated: Sticking/binding governor linkage or bent arm.

NO—Proceed to Fault Symptom **3660.**

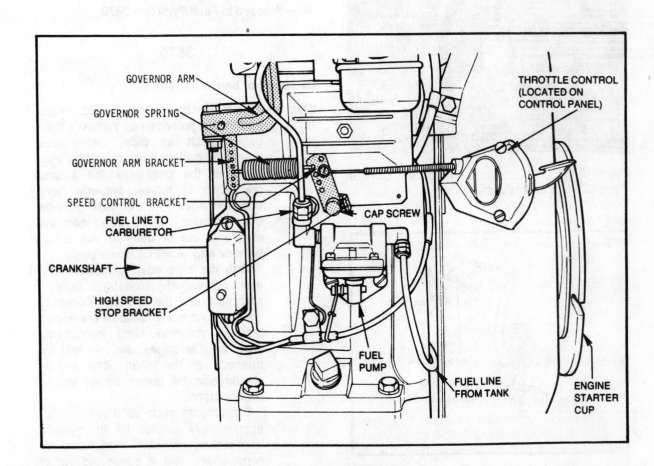

FIGURE 145   Typical Variable Speed Governor Arrangement

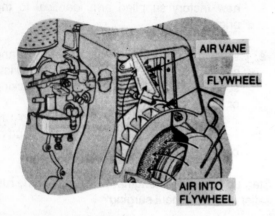

FIGURE 146 Typical Air Vane Type Governor

FIGURE 147 Governor Positive Control Type Linkage and Adjusting Screw

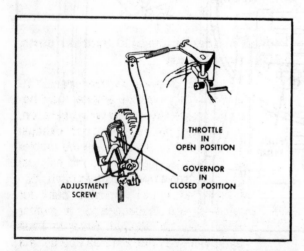

FIGURE 148 Typical Mechanical Flyweight Type Governor, Open Throttle

3660

Engine is loose

Step 1—Engine surge can be caused by the engine not being properly bolted down to the mower deck or chassis. If this condition exists, the engine THROTTLE controls will be vibrated about resulting in poor control. Shut down the engine and position controls to OFF. Remove the ignition cable from the spark plug and then check the engine mounting bolts to see that they are tight and secure. If not, tighten them. Replace ignition cable.

Start up and run the engine. Does the engine run satisfactorily, without surging or vibrating?

YES—Fault has been isolated: Engine was loose.

NO—Proceed to Fault Symptom **3670**.

3670

Damaged or bent crankshaft

Step 1—If the mower has seen very hard service and in the process has run over lawn debris such as pipe, heavy tree branches, garden hose couplings, etc., there is the possibility the engine crankshaft or mower blade(s) have been damaged. The crankshaft may be bent. Should this be true, then the engine is out of balance. An out of balance engine will cause surging.

Only the more obvious mower damage, damaged blade(s), pulleys, sprockets and the like can be easily checked. An out of balance crankshaft requires machine shop instruments such as dial gages, etc., to test for trueness of the shaft. The engine should also be tested on an engine service stand.

Equipments such as these are not economically justified for the general lawn mower owner. It is therefore recommended that a suspected out of balance crankshaft be checked by the Factory Authorized Service Dealer.

ENGINE RUNNING PROBLEMS EXHAUST NOISY, OR VERY QUIET, OR HEAVY SMOKING FAULT SYMPTOM 3700

Possible Causes:

- Defective muffler, noisy exhaust sounds ———— **See 3710**

- Clogged muffler or clogged exhaust ports,

  quiet exhaust sounds———————————— **See 3720**

- Carburetor improperly adjusted,

  causing smoking ————————————— **See 3730**

- Engine is burning oil, heavy smoking ———— **See 3740**

Initial Conditions Check List:

Perform all the pre-start checks given in the Check Lists of Fault Symptoms **1000** and **2000**.

3710

Defective muffler, noisy exhaust sounds

Step 1—Start up the engine and move the THROTTLE control back and forth from low to high and then back to low. As you move the control, listen to the engine sound. When the engine speeds up, the sound should get louder, then lower, when the speed is reduced. The engine sound at high speed should be a tolerable sound. The sound the engine makes should not be a very high roar, which can be heard a long distance away.

Does the engine sound become very loud to a roaring noise when the throttle is positioned to high speed?

YES—The fault has been isolated: Muffler is defective. Refer to Fault Symptom **3421** for replacement of the muffler.

NO—Fault has not been verified. Proceed to Fault Symptom **3720**.

3720

Clogged muffler or clogged exhaust ports, quiet exhaust sounds

Step 1—Two-cycle engines require special attention. This type of engine may become progressively quieter in operation, to the point where only a soft exhaust sound is heard. This means the muffler is plugged with carbon or the exhaust ports are clogged with carbon deposits.

Refer to Fault Symptom **3422** for removal and inspection of a muffler which may be clogged. Refer to Fault Symptoms **2611** or **2612** as appropriate (for horizontal or vertical mounted 2-cycle engine) for the procedure for clean out of carbon on clogged engine exhaust ports.

After muffler replacement or carbon clean out of exhaust ports, start up and run the engine.

Does the sound the engine makes appear to be normal, not loud nor very soft or quiet?

YES—Fault has been isolated.

NO—Refer to Fault Symptoms Index for other possible causes which can result in very quiet exhaust sounds.

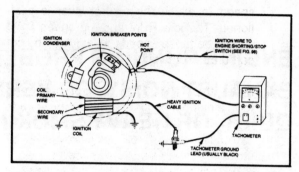

FIGURE 149 Tachometer Hook-up Connections

3730

Carburetor improperly adjusted, causing smoking

Step 1—When a mower engine is operating properly, there should be very little or no exhaust gas color of any kind. The exhaust gas should be colorless when the engine is operating in the midspeed range.

At both idle and wide open throttle, the exhaust should be barely noticeable. When the engine has a faulty carburetor, the exhaust gas will be very noticeable and smoky. The smoke can be dark gray to black in color—or even gray white in color.

If the fuel mixture is improperly adjusted, there can be a strong gasoline smell to the exhaust smoke.

Does the engine exhibit any of these exhaust signs or smells?

YES—Fault has been isolated. Proceed to Step 2.

NO—Proceed to Fault Symptom **3740.**

Step 2—If the carburetor is not adjusted properly, the engine will run rich and smoke excessively from the exhaust. Although an initial setting for the carburetor idle and high speed screws may be performed by hand, it will be necessary to use a Tachometer for final tuning. Connect up a Tachometer as shown in the accompanying Fig. 149. Proceed as follows:

a. The "hot point" is the point where the primary wire of the magneto coil is connected, along with the condenser wire, and the breaker points. It is from this common point that the hot lead wire goes to the starter switch or control cable short out location. The black or ground lead to the Tachometer can be connected to any convenient good ground point, but using the base of the spark plug is the preferred point.

b. Refer to the engine manufacturer's specification for the idle speed RPM and top speed RPM, or Service Procedure **4050.** If this information is not available, then a fair value for idle RPM speed is about 1300 or 1400 RPM. Top speed is about 3500 RPM.

Adjust the speed as follows:

c. Start up engine, let engine warm up at IDLE control lever position about 5 minutes, meanwhile read the RPM on the Tachometer.

d. Using your fingers, adjust the *idle* screw until the desired idle RPM is reached.

e. Reposition the control lever to maximum or high speed; let engine run at this speed a few minutes.

f. Using your fingers adjust the *power* screw until the desired RPM is reached.

g. Leave the Tachometer connected, and move the control back and forth from idle to high speed. Note that the engine should respond smoothly and should stay at the speed to which the control lever is positioned. The idle RPM and top speed RPM should be those speeds as adjusted by the carburetor screws.

h. Move the THROTTLE control back and forth a number of times to insure the speeds read out are those selected by the carburetor adjustments. If not, some fine tuning by small adjustments may be necessary.

The final adjustment can be performed by adjusting the high speed screw barely in or out until the engine accelerates smoothly but just barely gives off a very light exhaust smoke.

3740

Engine is burning oil, heavy smoking

Step 1—Should the engine be burning oil, the exhaust smoke will have a strong smell of burned oil. The odor is sharp and caustic. Proceed to Step 2 for special notation on heavy smoking of a 2-cycle engine.

Does the smoking exhaust have a sharp caustic smell?

YES—Fault has been verified: Engine burning oil. For 2-cycle engines, proceed to Step 2. For 4-cycle engines, refer to Service Procedure **4100**.

NO—Refer to Fault Symptom **3730** for proper carburetor adjustment. The fault has not been verified.

Step 2—Proper operation of 2-cycle engines requires that the fuel mixture of oil and gasoline follow the engine manufacturer's specification exactly. See Table I for some typical manufacturing mixtures. Check the fuel mixture which has been used and if there is any doubt on the mixture, drain the fuel tank completely and refill. Be sure to follow the manufacturer's specification. Start up and run the engine for about 10 minutes.

Does the engine exhaust give off heavy smoke?

YES—The fault has not been verified. Refer to Fault Symptom **2550** and perform the test of engine compression cited. If compression readings are low, engine is in need of mechanical overhaul. Refer to Service Procedure **4100**.

NO—Fault has been isolated: Too much oil in fuel mixture.

ENGINE RUNNING PROBLEMS
ENGINE MISFIRES, SKIPS
WHILE RUNNING

FAULT SYMPTOM 3800

Possible Causes:

- Water in fuel ——————————————————— **See 3810**
- Loose ignition cable on spark plug ——————— **See 3820**
- Faulty spark plug ——————————————— **See 3830**
- Faulty ignition breaker points or condenser ———— **See 3840**

Initial Conditions Check List:

a. Perform all the pre-start checks given in the Check Lists of Fault Symptoms **1000** and **2000**.

b. When an engine has been in service for a period of time, a problem may arise where the engine begins to misfire. Rather than constantly running smoothly, the engine will appear to hesitate, shudder, or skip a beat or two. The sound the engine makes will be interrupted each time the engine misfires. This sound will not be the usual buzzing sound, it will be a pop-pop sound separated by a very discernible quiet time span.

3810

Water in fuel

Step 1—Water in the fuel will cause misfire. Refer to Fault Symptom **3110** for a procedure to check and eliminate water in the fuel.

　　　After completion of the procedure, start up and run the engine.

Does the engine perform satisfactorily without any evidence of a misfire?

YES—Fault has been isolated: Water in fuel.

NO—Proceed to Fault Symptom **3820**.

3820

Loose ignition cable on spark plug

Step 1—A loose ignition cable connection can cause engine misfire. Check for a secure fit by the ignition cable cap end onto the spark plug. The cable cap should fit tightly. It should not be loose and bounce about on the plug terminal. If this is the case, tighten the cable cap by squeezing the outside of the cap with a pair of smooth jaw pliers. Proceed as follows:

a. Shut down the engine. Place control lever to OFF or STOP.

b. Tighten the cable cap end by squeezing very carefully on the outside of the cap with smooth jaw pliers.

c. Check the fit of the cable cap end onto the spark plug terminal. The fit should be tight and secure. Do not make cap too tight or it will no longer fit the terminal.
   Start up and run the engine.

Does the engine run satisfactorily without misfire?

YES—Fault has been isolated: Loose ignition cable on plug.

NO—Proceed to Fault Symptom **3830.**

3830

Faulty spark plug

Step 1—Too wide a setting of the plug gap will result in engine misfire. Proceed as follows:

a. Shut down engine. Position throttle lever to OFF.

b. Remove spark plug, measure gap and regap the plug as shown in Service Procedure **4004.** Note that the gap will vary with the manufacturer, and whether the engine is a 2- or 4-cycle engine. Consult your engine manufacturer's specification for

proper gap setting. If this information is not available, a general setting of .028 inches may be used.

c. While checking for proper gap setting, be certain to check that the plug type you are using is for your engine. Use of a wrong spark plug will also cause misfiring.

d. Clean the spark plug per Service Procedure **4004** before regapping.

e. Reinstall the plug and connect up the ignition cable.

Start up and run the engine.

Does the engine run satisfactorily without misfire?

YES—Fault has been isolated: Faulty spark plug.

NO—Proceed to Fault Symptom **3840.**

3840

Faulty ignition breaker points or condenser

Step 1—As the engine is used the breaker points in the ignition system are required to open and close and regulate the firing of the spark plug. Although a fairly long period of trouble-free operation is usually the case with breaker points, eventually they become pitted, corroded or dirty. As a result the spark plug voltage begins to deteriorate and insufficient voltage is provided the plug. Refer to Fig. 150 for sketches of ignition breaker points. Fig. 151 shows a set of breaker points installed in an engine. If the voltage begins to vary, engine misfiring can occur.

Another part in the ignition system which works in conjunction with the breaker points is the condenser. If the condenser begins to weaken by possible loss of electrical charging capability or breakdown of its insulation, or becomes leaky, then this will hasten and/or cause the breaker points to pit or corrode.

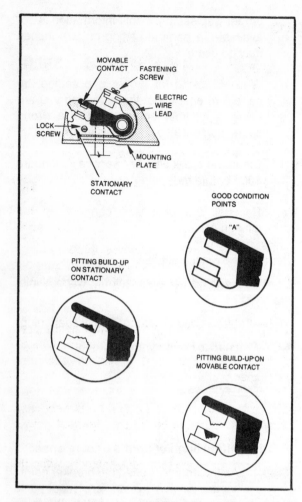

FIGURE 150 Good Condition Breaker Points and Poor Condition Pitted Points

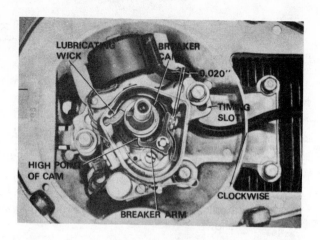

FIGURE 151 Typical Set of Ignition Breaker Points Installed in Engine

Finally the points must open and close at the proper time. If too soon or too late, due to the wrong gap, or a worn cam, the engine can misfire. Ignition timing is most critical in preventing engine misfiring. Proceed to Step 2.

Step 2—A quick determination can be made on the overall quality of the ignition breaker points and the condenser by the use of a test plug. Refer to Fault Symptom **3230,** Step 4, and the procedure for testing the magneto-electrical system. Perform the test cited.

Upon completion of the test was the spark produced on the special test plug satisfactory?

YES—The fault is not verified. A possible cause is improper tuning. Refer to Service Procedure **4030.**

NO—Fault has been isolated: Faulty ignition system components. Refer to Service Procedure **4030** for system overhaul.

ENGINE RUNNING PROBLEMS
ENGINE BACKFIRES

FAULT SYMPTOM   3900

Possible Causes:

- Ignition system defects ——————————— **See 3911**

- Sticking exhaust valve—4-cycle engine ———— **See 3912**

Initial Conditions Check List:

Perform all the pre-start checks given in the Check Lists of Fault Symptoms **1000** and **2000**.

3911

Ignition system defects

Step 1—Engine backfiring is related to misfiring. Backfiring is easily recognized by a sharp bang or loud claplike noise. The noise is actually some of the fuel-air mixture burning outside of the cylinder. Backfiring can be dangerous if ignored, since the muffler can be damaged or blown off the engine.

Backfiring can occur when the spark is delayed too long and as a result some of the combustion takes place outside of the cylinder. The prime causes of spark delay are related to improper timing. All of the ignition system components should be checked. To do so requires removal of the magneto flywheel in some engine designs. Refer to Service Procedure **4030—4030-3** if your mower has an internal power system:

a. Check the crankshaft key. If loose or worn, this can delay the spark and cause backfiring.

b. Check the keyway. If worn, this will require an oversized key. Use of the standard key in the worn keyway can delay the spark and cause backfiring.

c. Refer to Service Procedure **4030** starting with Step 1 for check of the other components of the ignition system as breaker points.

d. Check the timing of the breaker points per Procedure **4030**. Delayed point timing represents the most important possible cause of backfiring.

Upon completion of these ignition system checks /repairs and reassembly of the flywheel magneto, test run the engine.

Does the engine perform satisfactorily without misfiring?

YES—Fault isolated and verified.

NO—For a 4-cycle engine, proceed to Fault Symptom **3912**. For 2-cycle engine recheck Procedure **4030** for a sequence which may not have been performed or a component not checked.

3912

Sticking exhaust valve—4-cycle engine

Step 1—On 4-cycle engines a sticking exhaust valve can cause backfiring. If the valve delays in closing some of the fresh fuel air change can seep past the valve and ignite in the residue of the hot exhaust gases. Refer to Fault Symptom **2550** to isolate this fault.

Refer to Service Procedure **4100** for information on overhaul of intake/exhaust valves.

ENGINE RUNNING PROBLEMS
ENGINE CAME TO SUDDEN
ABRUPT STOP

FAULT SYMPTOM   3920

Possible Causes:

- Internal mechanical failure ————————— **See 3921**
- Wrong fuel mixture ——————————— **See 3922**
- Engine is overheating ————————— **See 3923**

Initial Conditions Check List:

Perform all the pre-start checks and conditions given in Fault Symptoms **1000** and **2000**.

3921

Internal mechanical failure

Step 1—Position THROTTLE control to OFF or STOP. Permit the engine to cool down to the same temperature as the surrounding air. After the engine has cooled, restart engine, be sure to use CHOKE control (if choke control is a separate control lever).

Did the engine start up and run with no apparent problems?

YES—Shut down engine and proceed to Fault Symptom **3922**.

NO—Proceed to Step 2.

Step 2—Position the THROTTLE control back to OFF. Slowly activate the starter in attempt to crank over the engine.

Does the engine crank (or turn over)?

YES—Proceed to Fault Symptom **3922**.

NO—If the engine does not crank over, it is quite probable that:

a.   The piston has seized (become frozen inside the cylinder).

b.   The connecting rod may have broken.

c.   The cylinder walls have become galled causing the piston to hang up in the cylinder.

d.   It is also possible the piston may have warped and changed size leading to its jamming in the cylinder.

For any of these possibilities, the engine is in need of major overhaul and repair. Refer to Fault Indication **4100**.

3922

Wrong fuel mixture

Step 1—Examine the container from which fuel is being used, and compare the fuel with the manufacturer's recommendations. For 2-cycle engines it is critical that the correct amount of oil be added in the mix.

Is the fuel mixture in accordance with the manufacturer's recommendation?

YES—If there is any doubt about what the fuel is in the mower fuel tank (or the mix in the tank), drain the tank completely, and refill with the recommended fuel. Proceed to Step 2.

NO—Drain the fuel tank and refill with the recommended fuel. Then proceed to Step 2.

Step 2—Restart the engine. Let the engine run for 10 to 15 minutes.

Does the engine run satisfactorily?

YES—Fault isolated and verified: Wrong fuel mixture.

NO—Proceed to Fault Symptom **3923**.

3923

Engine is overheating

Step 1—It is possible that the engine may be overheating. Examine the engine for signs of:
- Paint blistering, peeling or flaking off.
- For the paint getting much darker in color.
- Signs of smoke.

All, or combinations of these signs, are indications that the engine is overheating.

Does the engine exhibit any or several of these indications of overheating?

YES—Proceed to Step 2.

NO—The engine may be running hot for short periods due to very lean fuel adjustment. Refer to Fault Symptoms **3140** and **3330** for details on carburetor adjustment.

Step 2—In the case where any of these symptoms have been noticed, proceed as follows:

a. Clean away all dirt, grass clippings or grease from all parts of the engine.

b. Be particularly careful to clean out all of the fins around the engine cylinder. Dirt accumulations on the fins act as an insulation and prevent the proper transfer of heat away from the engine.

c. Be sure that shrouds, coverings, and particularly blower housings have been kept intact on the engine and that they have not been removed and left off. If the blower housing has been removed and left off the engine, then the cooling fins are not circulating the air properly and do not direct air over the fins. This will cause the engine to overheat. Continued overheating will cause internal engine damage which will render the engine useless.

   After cleaning away all dirt, check to be certain all necessary shrouds, blower housings and the like, have been re-installed. Start up and run the engine.

Does the engine run well with no indication of overheating?

YES—Fault has been isolated: Engine overheating.

NO—It is possible overheating is occurring because of an overly lean fuel mixture. This can be caused by air leaks into the engine crankcase. In 2-cycle engines the wrong fuel mixture can cause overheating. Refer to Fault Symptom **3300**.

ENGINE RUNNING PROBLEM
ENGINE VIBRATES BADLY

FAULT SYMPTOM   3930

Possible Causes:

- Engine loose ———————————————— **See 3931**
- Deck is cracked ———————————————— **See 3932**
- Shaft, blade or both are damaged ——————— **See 3933**

Initial Conditions Check List:

Perform all the pre-start checks given in Check Lists of Fault Symptoms **1000** and **2000**.

3931

Engine loose

Step 1—The engine should be mounted securely to the mower deck or chassis. If the bolts which hold the engine down are loose, or if some may be missing, the engine will move around relative to the deck and vibrate. When this occurs, poor THROTTLE control can result since the control is being bounced around on the engine.

a. Shut down the engine.

b. Remove the ignition cable from the spark plug.

c. Check for loose or missing hold down bolts. Tighten these bolts, if loose. Replace missing bolts.

Start up and run engine. Does the engine run satisfactorily without vibrating badly?

YES—Fault has been isolated: Loose/missing engine mounting bolts.

NO—Proceed to **3932**.

3932

Deck is cracked

Step 1—Examine the deck for cracks.

Are there any cracks?

YES—Potential fault has been isolated. Proceed to Step 2.

NO—Proceed to **3933**.

Step 2—Repair or replace deck. After repair start up engine.

Has the vibration disappeared?

YES—Fault isolated and repaired.

NO—Proceed to **3933**.

3933

Shaft, blade or ball are damaged

If the mower has seen very hard service and in the process has run over lawn debris, such as pipe, heavy tree branches, garden hose, couplings, etc., there is the possibility the engine crankshaft or mower blade(s) have been damaged. If the damage in turn, put a severe load on the crankshaft, the crankshaft may be bent. Should this be true, then the engine is out of balance. An out of balance engine will cause vibration.

Only the more obvious mower damage, damaged blade(s), pulleys, sprockets and the like can be easily checked. An out of balance crankshaft requires machine shop instruments such as dial gages, etc., to test for trueness of the shaft. The engine should also be tested on an engine service stand. Equipments such as these are not economically justified for the general lawn mower owner. It is therefore recommended that a suspected out of balance engine be checked by the Factory Authorized Service Dealer.

SERVICE PROCEDURES 4000

SERVICE PROCEDURES 4001

Exploded Line Drawing, Typical Lawn Mowers

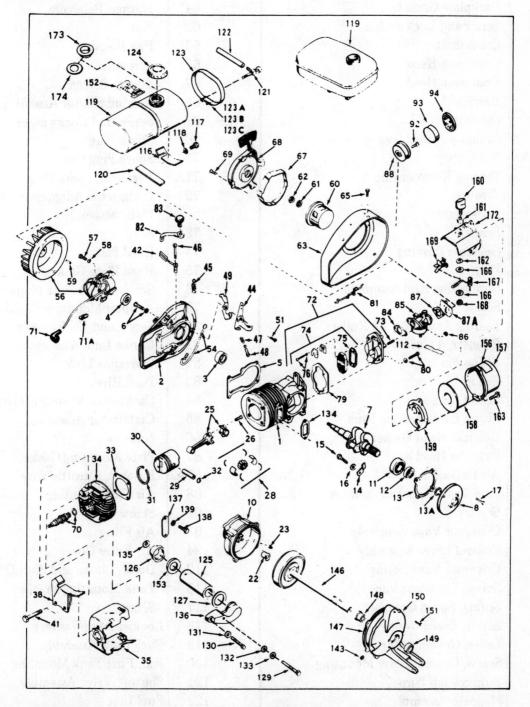

FIGURE 152   Jacobsen 321, Horizontal Engine, Exploded View

NOMENCLATURE LIST

Fig - Item	Description	Fig - Item	Description
152 - 1	Cylinder and Crankcase	152 - 57	Screw and Lockwasher
2	Backplate w/Bearing and Seal	58	Washer
3	Backplate Bearing	59	Flywheel Key
4	Backplate Oil Seal	60	Rewind Starter Hub
5	Backplate Gasket	61	Washer, Belleville
6	Screw and Lockwasher	62	Nut
7	Crankshaft	63	Fan Housing
8	Crankcase Head	65	Screw
10	Crankcase Head	67	Starter Screen
11	Bearing	68	Rewind Starter Assembly
12	Oil Seal	69	Screw and Lockwasher
13	Crankcase Head Gasket	70	Spark Plug
13A	Ring, Snap	71	Spark Plug Cap
14	Bearing Retainer Clip	71A	Terminal, Spark Plug
15	Screw	72	Carburetor Adapter and Reed
16	Lockwasher		Plate Assembly
17	Screw	73	Stud
22	Gearcase Bushing	74	Reed Plate
23	Plug	75	Reed Plate Gasket
25	Connecting Rod Assembly	76	Screw, Reed Plate Mounting
26	Screw	79	Adapter Mounting Gasket
28	Bearing Assembly, 28 Rollers,	80	Screw and Lockwasher
	2 Liners, 4 Guides	81	Choke Link Assembly
29	Piston Pin	82	Governor Link
30	Piston	83	Push Rivet
31	Piston Ring	84	Carburetor Mounting Gasket
32	Piston Pin Retaining Ring	85	Carburetor Assembly
33	Cylinder Head Gasket	86	Nut
34	Cylinder Head	87	Plate Mounting Gasket
35	Air Deflector	87A	Bracket, Throttle Wire
38	Fuel Tank Mounting Bracket	88	Air Filter Housing
41	Screw	92	Screw
42	Governor Vane Assembly	93	Air Filter
44	Control Lever Assembly	94	Air Filter Cover
45	Governor Vane Spring	112	Decal, Choke and Shut-Off
46	Screw, Governor Vane	116	Tank Mounting Bracket
47	Spring, Speed Control	117	Screw
48	Screw, Speed Control	118	Lockwasher (Washer)
49	Lever, Governor Spring	119	Fuel Tank Assembly
51	Screw, Control Lever Mounting	120	Pad, Fuel Tank Mounting
54	Stop Switch Wire	121	Shutoff Valve Assembly
56	Magneto Assembly	122	Fuel Line

NOMENCLATURE LIST (CONTD.)

Fig - Item	Description	Fig - Item	Description
152 -123	Tank Mounting Strap	152 - 147	Gear Reducer Cover Gasket
123A	Screw	148	Gear Reducer Cover Bearing
123B	Lockwasher	149	Gear Reducer Cover Seal
123C	Nut	150	Gear Reducer Cover
124	Fuel Tank Cap Assembly	151	Screw
125	Muffler	152	Decal, Name, Mix
126	Muffler Cap	153	Muffler Gasket
127	Muffler Mounting Gasket	156	Decal, Air Cleaner
128	Bolt	157	Air Filter Body
129	Screw	158	Air Filter
130	Screw	159	Bracket
131	Lockwasher	160	Knob
132	Muffler Head Gasket	161	Spacer
133	Washer	162	Washer
134	Exhaust Flange Gasket	163	Screw
135	Lockwasher	166	Washer
136	Muffler Head	167	Spring
137	Exhaust Flange Cover	168	Nut
138	Screw	169	Stop Switch Assembly
139	Lockwasher	172	Bracket Assembly
143	Plug	173	Baffle Assembly
146	Gear and Shaft Assembly	174	Gasket, Fuel Tank Cap

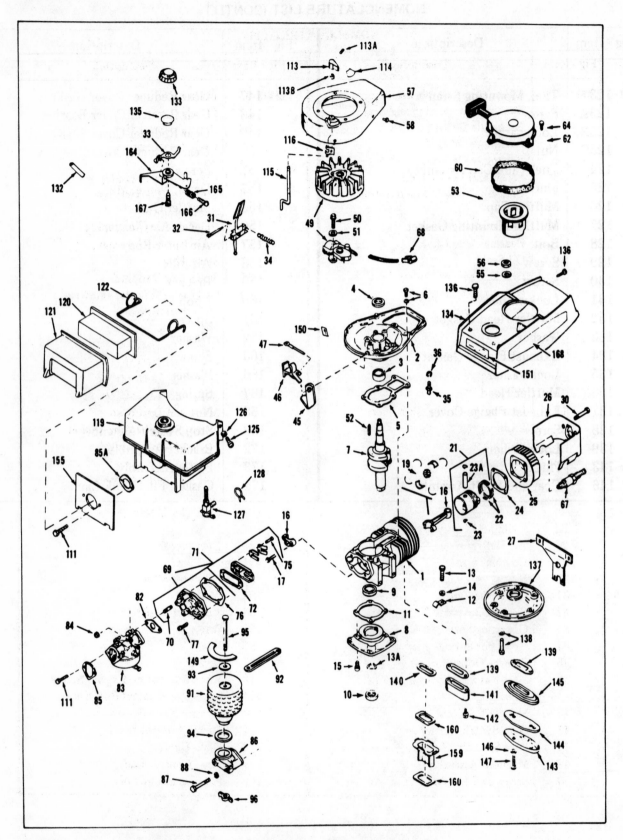

FIGURE 153   Jacobsen, 321, Vertical Engine, Exploded View

NOMENCLATURE LIST

Fig - Item	Description	Fig - Item	Description
153 - 1	Cylinder and Crankcase	153 - 51	Washer
2	Backplate w/Bearing and Seal	52	Flywheel Key
3	Backplate Bearing	53	Rewind Starter Hub
4	Backplate Oil Seal	55	Washer - Belleville
5	Backplate Gasket	56	Nut
6	Screw and Lockwasher	57	Fan Housing Assembly
7	Crankshaft, Including Reference	58	Screw and Lockwasher
	13A	60	Starter Screen
8	Crankcase Head	62	Rewind Starter Assembly
9	Bearing, Crankshaft	64	Screw and Lockwasher
10	Oil Seal	67	Spark Plug
11	Crankcase Head Gasket	68	Spark Plug Terminal
12	Bearing Retainer Clip	69	Carburetor Adapter and Reed
13	Screw		Plate Assembly
13A	Ring - Snap	70	Stud
14	Lockwasher	71	Reed Plate Assembly
15	Screw	72	Reed Plate Gasket
16	Connecting Rod Assembly	75	Screw and Lockwasher
17	Screw	76	Adapter Mounting Gasket
19	Bearing Assembly, 28 Rollers,	77	Screw and Lockwasher
	2 Liners and 4 Guides	82	Carburetor Mounting Gasket
21	Piston and Pin w/Rings	83	Carburetor Assembly
22	Piston Ring	84	Nut
23	Piston Pin Retaining Ring	85	Elbow Mounting Gasket
23A	Piston Pin	85A	Gasket
24	Cylinder Head Gasket	86	Air Filter Elbow
25	Cylinder Head	87	Screw
26	Air Deflector	88	Lockwasher
27	Cowling Brace	91	Air Filter
30	Screw	92	Air Filter Strap
31	Governor Vane	93	Washer
32	Screw - Governor Vane	94	Air Filter Gasket
33	Governor Spring Lever	95	Screw
34	Governor Spring	96	Wing Nut
35	Screw - Governor Detent	111	Screw
36	Lockwasher - Governor Lever	113	Knob-Control Rod w/Set Screw
	Mounting	113A	Set Screw, Control Rod Knob
45	Shut Off Switch	113B	Control Knob Spacer
46	Stop Switch Assembly	114	Decal - Control Panel
47	Wire Shut Off Switch	115	Rod - Control
49	Magneto Assembly	116	Clip - Control Rod
50	Screw and Lockwasher	119	Fuel Tank Assembly

NOMENCLATURE LIST (CONTD.)

Fig - Item	Description	Fig - Item	Description
153 -120	Air Filter	153 - 142	Screw and Lockwasher
121	Air Filter Cover	143	Muffler Cover
122	Air Filter Cover Spring	144	Muffler Baffle Plate
125	Screw	145	Muffler Body
126	Lockwasher	146	Washer
127	Shut Off Valve	147	Screw
128	Shut Off Valve Clamp	149	Decal - Air Filter
132	Fuel Line - 3-7/8" Long	150	Decal - Stop
133	Fuel Tank Cap	151	Decal - Fuel, Mix
134	Engine Cowling	155	Plate - Air Cleaner
135	Baffle Assembly	159	Heat Deflector
136	Screw	160	Gasket
137	Muffler Cover Lower Unit	164	Control Lever Assembly
138	Screw and Lockwasher	165	Spring-Speed Control
139	Exhaust Flange Gasket	166	Screw-Speed Control
140	Exhaust Flange Gasket	167	Screw-Control Lever Mounting
141	Exhaust Manifold	168	Decal - Name

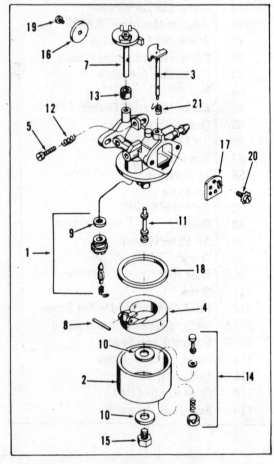

NOMENCLATURE LIST (FIXED JET)

Fig - Item	Description
154 - 1	Matched, Float Valve Seat, Spring and Gasket Assembly
2	Bowl Assembly - Float Bowl
3	Shaft Assembly - Choke
4	Float Assembly
5	Screw - Throttle Adjustment
7	Shaft Assembly - Throttle
8	Shaft - Float
9	Gasket - Float Valve Seat
10	Gasket - Nut To Bowl
11	Main Metering Nozzle
12	Spring Throttle Adjustment
13	Seal - Throttle Shaft
14	Bowl Drain Assembly
15	Retainer Screw
16	Throttle Plate
17	Choke Plate
18	Gasket - Bowl To Body
19	Screw
20	Screw
21	Spring - Choke Return

FIGURE 154   Carburetor, Exploded View

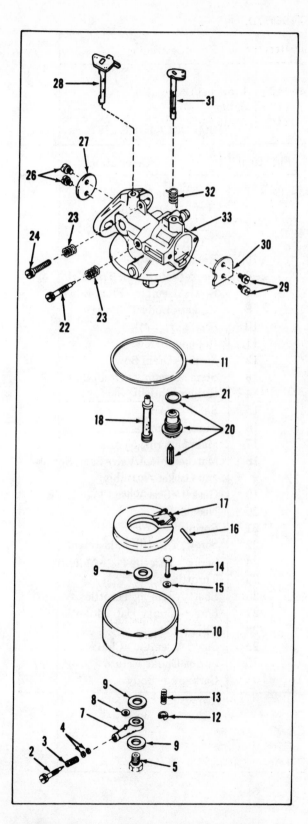

Fig - Item	Description
155 - 1	Carburetor Assembly
2	High-Speed Needle
3	Spring
4	"O" Rings
5	Retainer - Bowl
7	High-Speed Needle Housing
8	Rubber Gasket
9	Gasket - Bowl Nut To Bowl
10	Bowl Assembly - Float Bowl
11	Gasket - Body To Bowl
12	Retainer Screw
13	Spring - Drain Bowl
14	Stem Assembly - Drain Bowl
15	Rubber Gasket
16	Shaft - Float
17	Float Assembly
18	Main Metering Nozzle
20	Matched Float Valve, Seat, Spring and Gasket Assembly
21	Gasket - Seal Valve Float
22	Needle - Idle
23	Spring - Throttle Adjustment Screw
24	Screw - Idle Speed
26	Screw - Throttle Plate Mounting
27	Throttle Plate
28	Shaft Assembly - Throttle
29	Screw - Choke Plate Mounting
30	Choke Plate
31	Shaft Assembly - Choke
32	Choke Return Spring
33	Carburetor Body

FIGURE 155   Carburetor, Exploded View

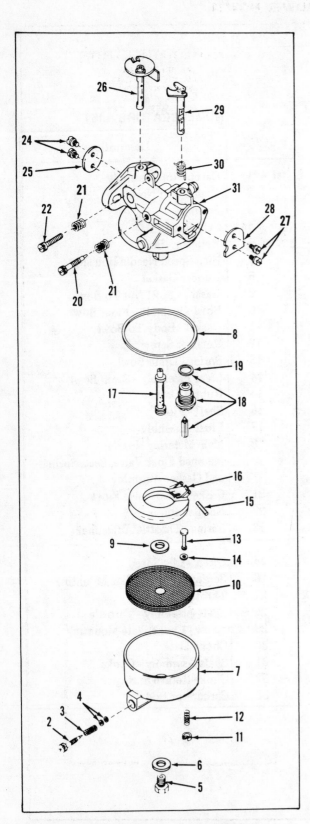

NOMENCLATURE LIST

Fig - Item	Description
156 - 1	Carburetor Assembly
2	High-Speed Needle
3	Spring
4	"O" Rings
5	Retainer Screw
6	Gasket - Bowl Nut To Bowl
7	Bowl Assembly - Float Bowl
8	Gasket - Body To Bowl
10	Screen
11	Retainer Screw
12	Spring - Drain Bowl
13	Stem Assembly - Drain Bowl
14	Rubber Gasket
15	Shaft - Float
16	Float Assembly
17	Main Metering Nozzle
18	Matched Float Valve Seat, Spring and Gasket Assembly
19	Gasket - Seal Valve Float
20	Needle, Idle
21	Spring
22	Screw, Throttle Adjustment
24	Screw - Throttle Plate Mounting
25	Throttle Plate
26	Shaft Assembly Throttle
27	Screw - Choke Plate Mounting
28	Choke Plate
29	Shaft Assembly - Choke
30	Spring Choke Return
31	Carburetor Body

Figure 123. Carburetor Exploded View.
FIGURE 156 Carburetor, Exploded View

20" ROTARY POWER MOWER

FOR PARTS LIST SEE FIGURE 158
ALL UNNUMBERED
PARTS INTER-CHANGEABLE
WITH OPPOSITE SIDE

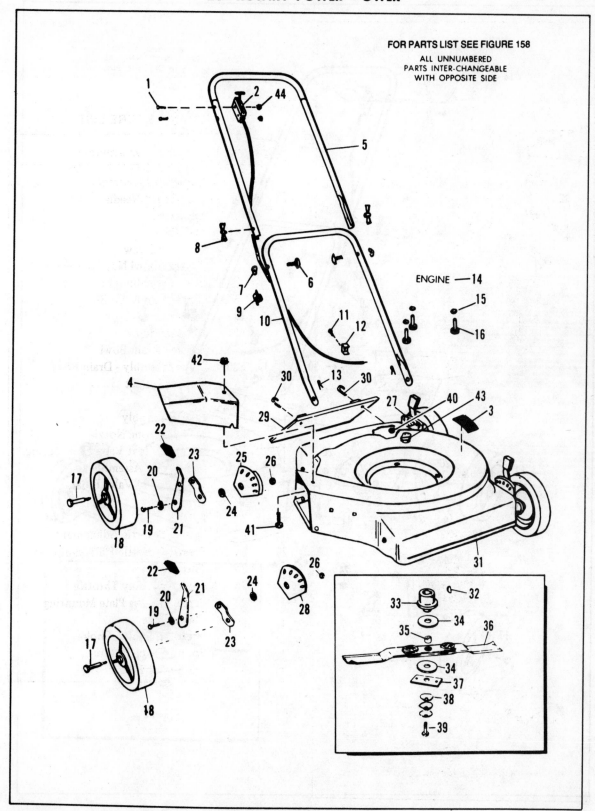

FIGURE 157   AMF   Rotary   Mower   Model 1250, Frame Exploded View

22" ROTARY POWER MOWER

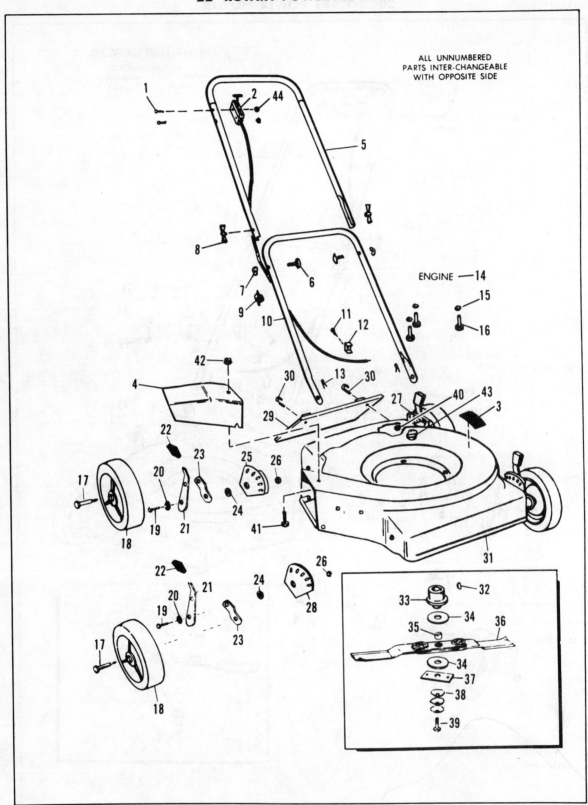

ALL UNNUMBERED
PARTS INTER-CHANGEABLE
WITH OPPOSITE SIDE

ENGINE

FIGURE 158   AMF Rotary Mower Model 1253, Frame Exploded View

20" ROTARY POWER MOWER

ALWAYS GIVE THE FOLLOWING INFORMATION WHEN ORDERING REPAIR PARTS:

1. THE PART NUMBER   2. THE PART NAME   3. QUANTITY DESIRED   4. THE MODEL NUMBER— **12500300**

Your Unit is Right Hand (R.H.) or Left Hand (L.H.) as you stand behind it.

DO NOT USE KEY NUMBERS WHEN ORDERING REPAIR PARTS, ALWAYS USE PART NUMBERS.

Key No.	Part No.	Description
1	9415101	*No. 10-24 x 1¾" Hex Head Screw
2	41110	Throttle Control
3	36664	Foot Pad
4	41284	Chute Deflector
5	40950	Upper Handle
6	35475	5/16"-18 x 1¾" Curv. Hd. Carr. Bolt
7	8728	Formed Washer
8	39592	Locking Knob
9	32951	Wire Clip
10	35787	Lower Handle
11	995363	*No. 10-32 x ⅝ Ind. Hd. S. T. Sc.
12	27521	Wire Clamp
13	121224	*3/32" x 1" Cotter Pin
14	ENGINE	See Footnote Below
15	120382	⅜" Spring Lock Washer
16	35326	⅜" 16 x 1⅛" Hex Hd. Taptite Sc.
17	27142	Axle Bolt
18	41346	Wheel & Tire Assembly
19	24246	Shoulder Bolt
20	32327	Wave Washer
21	36817	Height Control Lever
22	20252	Selector Knob

Key No.	Part No.	Description
23	36818	Pivot Arm
24	36821	Fiber Washer
25	36814	Handle Mount Bracket R.H.
26	9413534	3/18" - 16 Hex Locknut
27	36815	Handle Mount Bracket L.H.
28	36816	Height Control Bracket
29	41285	Toe Guard
30	41372	Bolt, Toe Guard
31	40947	Main Frame Assembly
32	2125	No. 505 Woodruff Key
33	13163	Shaft Adapter (Complete with all mating parts)
34	2277	Fiber Washer
35	20057	Bushing
36	41195	20" Blade
37	35388	Blade Adapter Washer
38	2483	Spring Washer
39	35343	⅜"-24 x 2¼" Hex Hd. Lock Sc.
40	41375	¼"-20 Hex Nut - Eslok
41	180016	*¼"-20 x ½" Hex Hd. Screw
42	997314	*¼"-20 Hex Locknut
43	40264	Wash-Out Port
44	997316	*No. 10-24 Hex Locknut

* Standard Hardware Items May Be Purchased Locally

AMF Western Tool Division, P.O. Box 377, Des Moines, Iowa 50302

Replacement engines and parts are obtainable from the Engine Manufacturer's authorized Service Stations who are also to be contacted in regards to the Engine Warranty. See your Engine Manual for location of these stations.

FIGURE 159   AMF Rotary Mower Model 1250, Nomenclature List

22" ROTARY POWER MOWER

ALWAYS GIVE THE FOLLOWING INFORMATION WHEN ORDERING REPAIR PARTS:

1. THE PART NUMBER   2. THE PART NAME   3. QUANTITY DESIRED   4. THE MODEL NUMBER— **12530200**

Your Unit is Right Hand (R.H.) or Left Hand (L.H.) as you stand behind it.

DO NOT USE KEY NUMBERS WHEN ORDERING REPAIR PARTS, ALWAYS USE PART NUMBERS.

Key No.	Part No.	Description
1	9415101	*No. 10-24 x 1¾" Hex Head Screw
2	41110	Throttle Control
3	36664	Foot Pad
4	41286	Chute Deflector
5	40950	Upper Handle
6	35475	5/16"-18 x 1¾" Curv. Hd. Carr. Bolt
7	8728	Formed Washer
8	39592	Locking Knob
9	32951	Wire Clip
10	41234	Lower Handle
11	995363	*No. 10-32 x ⅝ Ind. Hd. S. T. Sc.
12	27521	Wire Clamp
13	121224	*3/32" x 1" Cotter Pin
14	ENGINE	See Footnote Below
15	120382	⅜" Spring Lock Washer
16	35326	⅜" 16 x 1⅛" Hex Hd. Taptite Sc.
17	27142	Axle Bolt
18	41346	Wheel & Tire Assembly
19	24246	Shoulder Bolt
20	32327	Wave Washer
21	36817	Height Control Lever
22	20252	Selector Knob

Key No.	Part No.	Description
23	36818	Pivot Arm
24	36821	Fiber Washer
25	36814	Handle Mount Bracket R.H.
26	9413534	3/18" - 16 Hex Locknut
27	36815	Handle Mount Bracket L.H.
28	36816	Height Control Bracket
29	41287	Toe Guard
30	41372	Bolt, Toe Guard
31	41031	Main Frame Assembly
32	2125	No. 505 Woodruff Key
33	13163	Shaft Adapter (Complete with all mating parts)
34	2277	Fiber Washer
35	20057	Bushing
36	41193	22" Blade
37	35388	Blade Adapter Washer
38	2483	Spring Washer
39	35343	⅜"-24 x 2¼" Hex Hd. Lock Sc.
40	41375	¼"-20 Hex Nut - Eslok
41	180016	*¼"-20 x ½" Hex Hd. Screw
42	997314	*¼"-20 Hex Locknut
43	40264	Wash-Out Port
44	997316	*No. 10-24 Hex Locknut

* Standard Hardware Items May Be Purchased Locally

AMF Western Tool Division, P.O. Box 377, Des Moines, Iowa 50302

Replacement engines and parts are obtainable from the Engine Manufacturer's authorized Service Stations who are also to be contacted in regards to the Engine Warranty. See your Engine Manual for location of these stations.

FIGURE 160   AMF Rotary Mower Model 1253, Nomenclature List

18" ROTARY MOWER

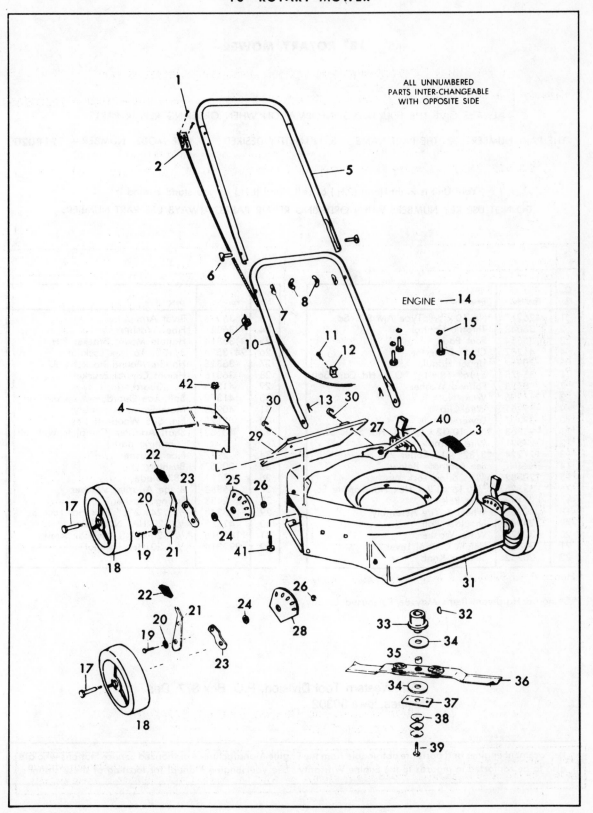

ALL UNNUMBERED
PARTS INTER-CHANGEABLE
WITH OPPOSITE SIDE

ENGINE — 14

FIGURE 161   AMF Rotary Mower Model 1218, Frame Exploded View

18" ROTARY MOWER

ALWAYS GIVE THE FOLLOWING INFORMATION WHEN ORDERING REPAIR PARTS:

1. THE PART NUMBER   2. THE PART NAME   3. QUANTITY DESIRED   4. THE MODEL NUMBER— **12180200**

Your Unit is Right Hand (R.H.) or Left Hand (L.H.) as you stand behind it.

DO NOT USE KEY NUMBERS WHEN ORDERING REPAIR PARTS, ALWAYS USE PART NUMBERS.

Key No.	Part No.	Description
1	9426217	*No. 10 x ½" Type AM SM. Sc.
2	36048	Throttle Control
3	36664	Foot Pad
4	41282	Chute Deflector
5	40952	Upper Handle
6	35475	5/16"-18 x 1¾" Curv. Hd. Carr. Bolt
7	8728	Formed Washer
8	997206	Wing Nut
9	32951	Wire Clip
10	39307	Lower Handle
11	995363	*No. 10-32 x ⅝ Ind. Hd. S. T. Sc.
12	27521	Wire Clamp
13	121224	*3/32" x 1" Cotter Pin
14	ENGINE	See Footnote Below
15	120382	⅜" Spring Lock Washer
16	35326	⅜" 16 x 1⅛" Hex Hd. Taptite Sc.
17	27142	Axle Bolt
18	40965	Wheel & Tire Assembly
19	24246	Shoulder Bolt
20	32327	Wave Washer
21	36817	Height Control Lever
22	20252	Selector Knob

Key No.	Part No.	Description
23	36818	Pivot Arm
24	36821	Fiber Washer
25	36814	Handle Mount Bracket R.H.
26	9413534	3/18"-16 Hex Locknut
27	36815	Handle Mount Bracket L.H.
28	36816	Height Control Bracket
29	41283	Toe Guard
30	41372	Bolt, Toe Guard
31	40955	Main Frame Assembly
32	2125	No. 505 Woodruff Key
33	13163	Shaft Adapter (Complete with all mating parts)
34	2277	Fiber Washer
35	20057	Bushing
36	41191	18" Blade
37	35388	Blade Adapter Washer
38	2483	Spring Washer
39	35343	⅜"-24 x 2¼" Hex Hd. Lock Sc.
40	41375	¼"-20 Hex Nut - Eslok
41	180016	*¼"-20 x ½" Hex Hd. Screw
42	997314	*¼"-20 Hex Locknut

* Standard Hardware Items May Be Purchased Locally

AMF Western Tool Division, P.O. Box 377, Des Moines, Iowa 50302

Replacement engines and parts are obtainable from the Engine Manufacturer's authorized Service Stations who are also to be contacted in regards to the Engine Warranty. See your Engine Manual for location of these stations.

FIGURE 162   AMF Rotary Mower Model 1218, Nomenclature List

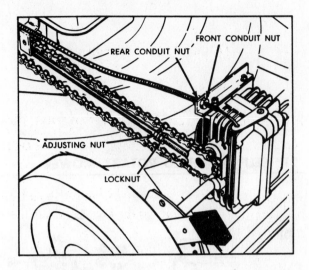

FIGURE 163   Self Propelled Mower Drive
Chain and Gear Box

DRIVE CHAIN ADJUSTMENT

After a period of use the drive chain will stretch. Remove the chain cover and fasteners. Loosen locknut from adjusting nut. (See Fig. 164.) Tighten adjusting nut one-half (½) turn at a time until proper chain tension is achieved which is ⅛"–¼" deflection. When adjustment is complete, tighten locknut and replace chain cover and fasteners.

GEARBOX NEUTRAL ADJUSTMENT

NOTE:

When instructions mention turning nut clockwise or counterclockwise this will be as viewed from the operator's position.

If when shifting from low gear into neutral, the machine fails to stop, turn the rear conduit nut counterclockwise one (1) turn, then the front conduit nut one (1) turn counterclockwise. If the unit still fails to shift from low into neutral keep repeating the above steps until desired condition is met.

If when shifting from high gear into neutral the unit fails to stop, adjust for neutral in the reverse order from the above, turning the front conduit nut clockwise one (1) turn, then the rear.

21" SELF PROPELLED ROTARY MOWER

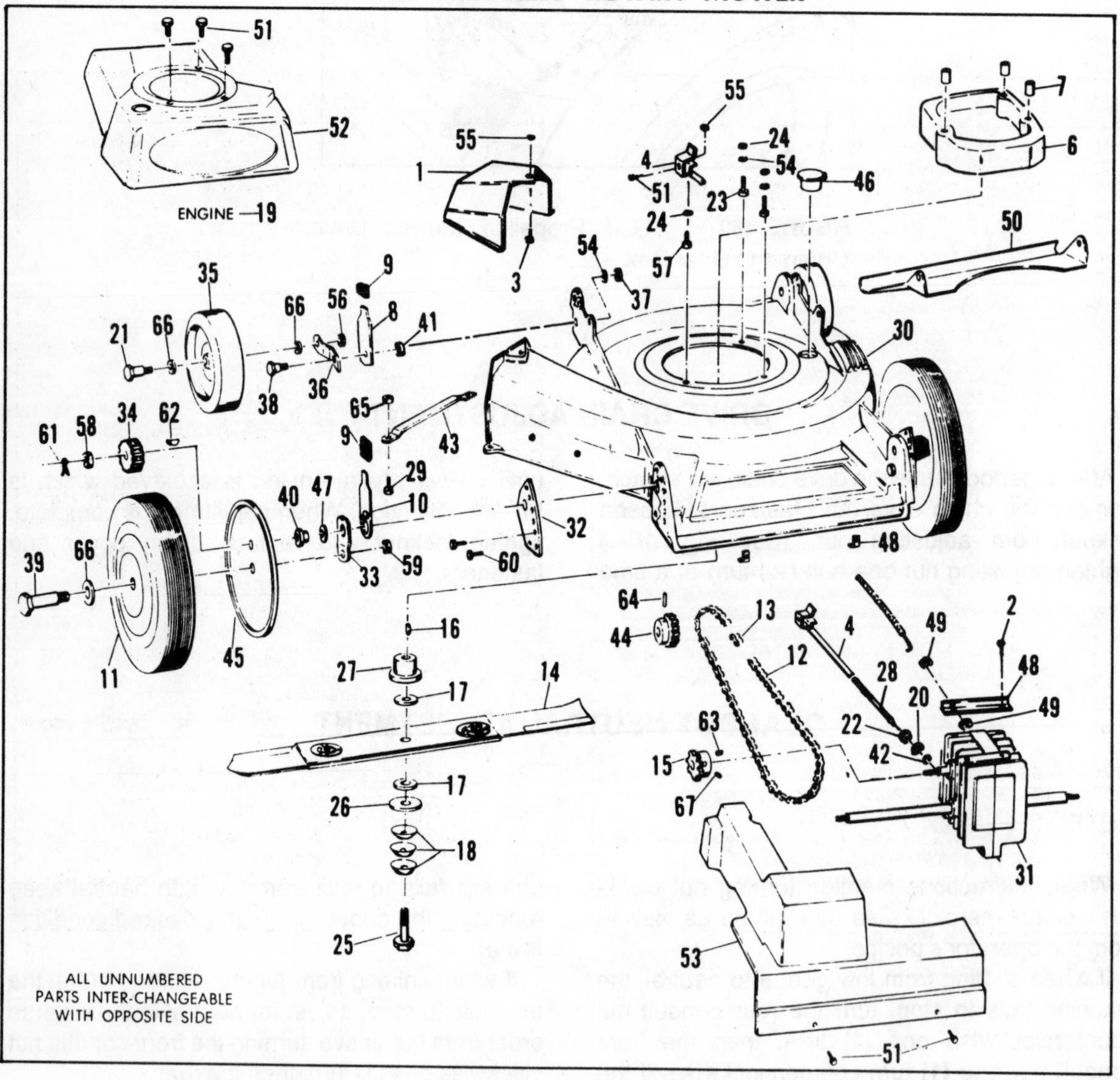

ENGINE — 19

ALL UNNUMBERED
PARTS INTER-CHANGEABLE
WITH OPPOSITE SIDE

FIGURE 164   AMF   Rotary   Mower   Model 1258 or 1259, Frame Exploded View

21" SELF PROPELLED ROTARY MOWER

ALWAYS GIVE THE FOLLOWING INFORMATION WHEN ORDERING REPAIR PARTS:

1. THE PART NUMBER   2. THE PART NAME   3. QUANTITY DESIRED   4. THE MODEL NUMBER—
12580100
or
12590100

Your Unit is Right Hand (R.H.) or Left Hand (L.H.) as you stand behind it.

DO NOT USE KEY NUMBERS WHEN ORDERING REPAIR PARTS, ALWAYS USE PART NUMBERS.

FIGURE 1 PARTS LIST FOR BOTH MODELS

Key No.	Part No.	Description	Key No.	Part No.	Description
1	41214	Chute Extension	35	36677	Wheel and Tire Assembly
2	35258	No. 10-24 x ⅜" Hex Hd. Taptite Sc.	36	40226	Wheel Adjusting Arm, Rear
3	126358	*5/16"-18 x 1" Carriage Bolt	37	40237	Handle Pivot Nut
4	41028	Brace Tube & Bracket Assy.	38	40238	Shoulder Bolt
5	40275	Leaf Mulcher Plate (Not Illustrated)	39	40239	Shoulder Bolt
6	39801	Engine Mount Assembly	40	40240	Flanged Radial Bearing
7	35308	Spacer, Engine Mount	41	39764	Washer
8	41222	Selector Spring, Rear	42	9417373	*No. 10 Washer
9	20252	Knob	43	40259	Chute Guard
10	41245	Selector Spring, Front	44	40261	Sprocket, 8 Tooth
11	40214	Wheel and Assembly	45	40263	Wheel Cover
12	40273	Chain Assembly	46	40264	Washout Plug
13	20304	Connecting Link	47	39766	Washer
14	40274	21" Flexor Blade	48	41029	Anchor Bracket, Clutch Cable
15	562	Sprocket, 9 Tooth	49	41242	Conduit Nut
16	2125	*No. 505 Woodruff Key	50	40271	Toe Guard
17	2277	Fiber Washer	51	9426213	No. 8 x ⅝" Hex Slot Washer Hd. Screw
18	2483	Spring Washer	52	36847	Engine Shroud
19	Engine	See Footnote Below	53	41249	Chain and Gear Box Shroud
20	271184	*5/16"-18 Keps Nut	54	120382	⅜" Spring Lock Washer
21	20753	Wheel Axle	55	9413447	*5/16"-18 Hex Locknut
22	124824	*5/16"-18 Hex Jam Nut	56	9413534	*⅜"-16 Hex Locknut
23	35304	⅜"-16 x 2½" Hex Hd. Taptite Sc.	57	122089	*5/16"-18 x 3" Hex Hd. Bolt
24	35336	⅜" Washer	58	125127	Hexagon Slotted Nut
25	35344	⅜"-20 x 2½" Hex Hd. Lock Screw	59	145027	⅜"-16 Locknut
26	35388	Blade Adapter Washer	60	39364	¼"-20 x ½" Pan Hd. Taptite Screw
27	13165	Shaft Adapter (Complete with all fasteners and Mating Parts)	61	121222	*3/32" x ¾" Cotter Pin
28	41025	Adjusting Rod	62	218218	No. 207 Woodruff Key
29	41281	¼"-20 x 1" Carriage Bolt	63	427261	No. 213 Woodruff Key
30	41215	21" Deck (Cast)	64	455481	3/16" x 1" Pin
31	41248	Gear Box Assy. (See Separate List)	65	997314	*¼"-20 Hex Locknut
32	40197	Wheel Selector Bracket	66	996425	*½" Washer
33	40198	Wheel Adjusting Arm, Front	67	998503	5/16"-18 x 5/16" Nylock Set Screw
34	40216	Drive Gear			

*Standard Hardware Items May Be Purchased Locally

Replacement engines and parts are obtainable from the Engine Manufacturer's authorized Service Stations who are also to be contacted in regard to the Engine Warranty. See your Engine Manual for location of these stations.

FIGURE 165   AMF Rotary Mower Model 1258 or 1259, Nomenclature List

21" SELF PROPELLED ROTARY MOWER

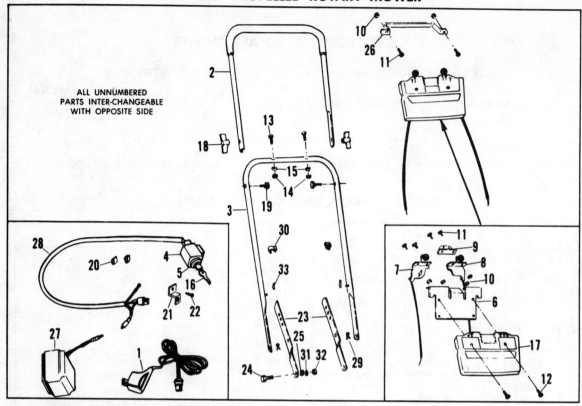

ALL UNNUMBERED
PARTS INTER-CHANGEABLE
WITH OPPOSITE SIDE

FIGURE 166   Parts List for AMF Model 12590100

Key No.	Part No.	Description
1	36655	Battery Charger
2	40249	Upper Handle Assembly
3	39359	Lower Handle Assembly
4	36133	Switch, Key Start
5	36135	Face Nut
6	40269	Control Mount Plate
7	40292	Throttle Control Assembly
8	40293	Gear Box Control Assembly
9	39776	Neutral Bracket
10	997316	*No. 10-24 Locknut
11	35291	No. 10-24 x ½" Hex Slot Hd. Screw
12	9426217	*No. 10 x ½" Washer Hd. Screw
13	121926	*¼"-20 x 1½" Hex Hd. Screw
14	997314	*¼"-20 Hex Locknut
15	8728	Formed Washer
16	36131	Key, Switch
17	40285	Control Panel

Key No.	Part No.	Description
18	39592	Locking Knob
19	35475	5/16"-18 x 1⅝" Curved Hd. Bolt
20	36132	Harness Clip
21	36267	Harness Bracket
22	36270	No. 8-23 x ⅝" Hex Hd. Screw
23	40248	Handle Link
24	120229	Shoulder Bolt
25	40252	Washer, Rubber
26	40282	Battery Bracket
27	40284	Battery
28	40290	Harness Assembly
29	121224	*3/32" x 1" Cotter Pin
30	32951	Cable Clip
31	446188	Washer
32	9413447	*5/16"-18 Hex Locknut
33	31448	Hair Pin Cotter

PARTS LIST FOR MODEL 12580100

All parts are the same as the list above except the following:

Key No.	Patt No.	Description
13	997314	¼"-20 Hex Locknut
17	40299	Control Panel

Key No.	Part No.	Description
Key No's. 1, 4, 5, 16, 20, 21, 22, 26, 27 & 28 not used.		

* Standard Hardware Items May Be Purchased Locally

FIGURE 167   Parts List for AMF Model 12580100

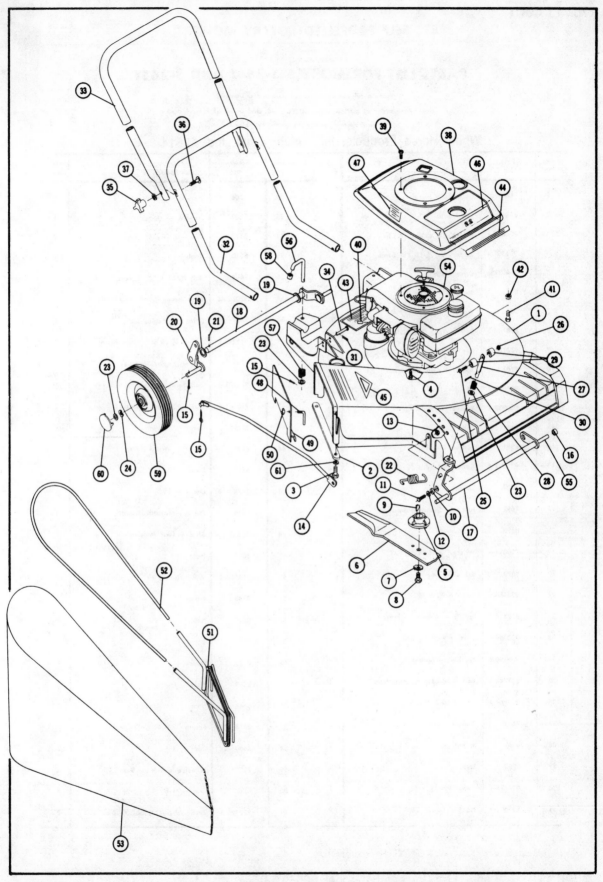

FIGURE 168   Wheel Horse Rotary Mower Models 3-2641, 3-2651, Exploded View

PARTS LIST FOR MODELS 3-2641 AND 3-2651

Parts available only through Authorized Dealers.

When ordering parts always list Part No. and name of Part.

(Specifications subject to change without notice.)

Wheel Horse Products, Inc. South Bend, Ind. 46614

Item No.	Part No.	Description	No. Req'd.	Item No.	Part No.	Description	No. Req'd.
1	8580	Deck	1	32	8608	Handle — Lower	1
2	8581	Toe Guard	1	33	8609	Handle — Upper	1
3	926695-4	Bolt 5/16-18 x 3/4 Self Tap	2	34	5523	Pin — Clevis 1/4 Dia.	2
4	911301-4	Screw Hex 5/16-18 x 1 Sems	3	35	8610	Knob	2
5	MW-7726	Drive Blade	1	36	8684	Bolt 5/16-18 x 1 3/4 Curved Head	2
6	8551	Blade	1	37	920155-4	Lockwasher 5/16 Internal Tooth	2
7	1336	Washer	1	38	7657	Shroud — Engine	1
8	908184-5	Bolt 3/8-24 x 1 1/4 Nylok	1	39	911793-4	Screw 1/4-28 x 1/2 Pan Head	4
9	937155	Key Hi-Pro	1	40	9752	Cap — Engine Throttle	1
10	9101	Spacer Height Adjustment	2	41	911781-4	Bolt Sems 1/4-20 x 1/2	1
11	908019-4	Bolt Hex 5/16-18 x 1	2	42	915661-4	Nut 1/4-20 Elastic Stop	1
12	920038-4	Washer — Plain 5/16	2	43	8619	Decal — Caution	1
13	915112-6	Nut 5/16-18	2	44	8620	Decal — Wheel Horse	1
14	8591	Rod — Wheel	1	45	5699	Decal — ASA Triangle	1
15	932008-4	Cotter Pin 3/32 x 3/4	8	46	8875	Decal — 22"	1
16	6309	Spacer	1	47	9936	Decal — Throttle Control	1
17	8592	Height Adjustment Belcrank	1	48	8624	Stud — Leaf Mulcher	2
18	8595	Axle — Rear	1	49	8625	Plate	1
19	920013-4	Washer — Plain 5/8 SAE	2	50	915584-4	Wing Nut 1/4-20	2
20	8598	Belcrank Rear	1	51	9183	Frame	1
21	933186	Roll Pin 3/16 x 7/8	1	52	8989	Hoop	1
22	8594	Spring — Counter Balance	1	53	8983	Bag	1
23	920009-4	Washer Plain 3/8 SAE	8	54	8582	Engine	1
24	MW-4405	Washer — Spring 3/8	4	55	8589	Axle — Front	1
25	911743-4	Bolt #10-24 x 3/4 Pan Head	1	56	8615	Hook	2
26	915964-4	Nut #10-24 Keps	1	57	8616	Spring	2
27	8601	Plunger	1	58	3680	Cap	2
28	8602	Spring	1	59	7159	Wheel and Tire — 8 x 1.75	4
29	8982	Knob	2	60	6312	Cap — Hub	4
30	934786-4	Cotter Pin 1/8 x 3/4	1	61	920123-4	Lockwasher External 5/16 Dia.	2
31	933505-4	Hairpin Cotter	2				

FIGURE 169   Wheel Horse Rotary Mower Models 3-2641, 3-2651, Nomenclature List

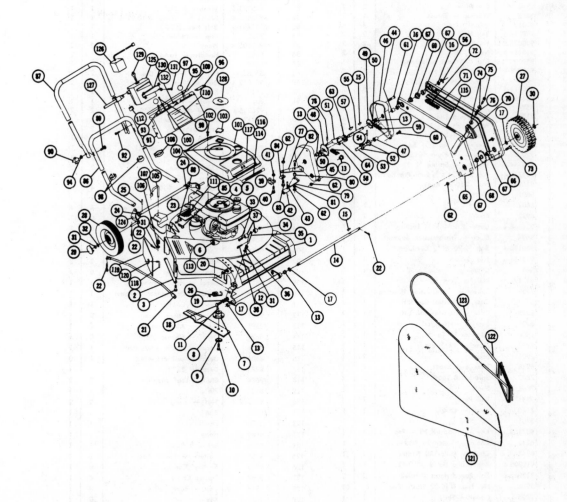

FIGURE 170   Wheel Horse Rotary Mower Models 3-1641, 3-1651, 3-1741, 3-1751, Exploded View

MODELS 3-1641 & 3-1651 (RECOIL) AND MODELS 3-1741 & 3-1751 (ELECTRIC)

When ordering parts always list Part No. and Name of Part.

Parts available only through Authorized Dealers.

(Specifications subject to change without notice.)

Ref. No.	Part No.	Description	Req'd. Recoil	Req'd. Electric
1	8580	Deck	1	1
2	8581	Toe Guard	1	1
3	926695-4	Bolt 5/16-18 x 3/4 Self Tap	2	2
4	8583	Engine	1	0
5	8584	Engine	0	1
6	911301-4	Screw Hex 5/16-18 x 1 Sems	3	3
7	MW-7726	Drive Blade	1	1
8	8551	Blade	1	1
9	1336	Washer	1	1
10	908184-5	Bolt 3/8-24 x 1 1/4 Nylok	1	1
11	937155	Key Hi — Pro	1	1
12	8585	Height Adjustment — Belcrank	1	1
13	6230	Bearing — Flanged 1/2 I.D.	5	5
14	100322	Axle — Front	1	1
15	9380	Key Hi — Pro	2	2
16	3765	Washer — Thrust	2	2
17	9101	Spacer	4	4
18	908019-4	Bolt Hex 5/16-18 x 1	2	2
19	920038-4	Washer — Plain 5/16 U.S.	2	2
20	915112-6	Nut 5/16-18	2	2
21	8591	Rod — Wheel Adjustment	1	1
22	932008-4	Cotter Pin 3/32 x 3/4	8	8
23	8595	Axle — Rear	1	1
24	920013-4	Washer — Plain SAE	2	2
25	933186	Roll Pin 3/16 x 7/8	1	1
26	8594	Spring — Counter Balance	1	1
27	7158	Wheel and Tire 8 x 1.75	2	2
28	7159	Wheel and Tire 8 x 1.75	2	2
29	6312	Cap — Hub	4	4
30	920011-4	Washer — Plain 1/2 SAE	2	2
31	920009-4	Washer — Plain 3/8 SAE	5	5
32	MW-4405	Washer — Spring 3/8	2	2
33	911743-4	Bolt #10-24 x 3/4 Sems Pan Head	1	1
34	915964-4	Nut #10-24 Keps	1	1
35	8601	Plunger	1	1
36	8602	Spring	1	1
37	8982	Knob	2	2
38	934768-4	Cotter Pin 1/8 x 3/4 Locking	1	1
39	8604	Block — Transmission Support	1	1
40	900196-4	Bolt 1/4-20 x 3/4 Carriage	1	1
41	915111-6	Nut 1/4-20	1	1
42	8605	Coupling	1	1
43	933197	Roll Pin 7/32 x 3/4	1	1
44	7131	Spacer — Transmission Housing	1	1
45	6214	Case — Transmission R.H.	1	1
46	6215	Case — Transmission L.H.	1	1
47	6229	Bearing — Flanged 1/2 I.D.	1	1
48	7900	Sprocket 19 Tooth	1	1
49	7903	Sprocket 12 Tooth	1	1
50	7901	Bearing	2	2
51	6238	Yoke and Shaft	1	1
52	7863	Sprocket 8 Tooth	2	2
53	7862	Shaft	1	1
54	933171	Roll Pin	2	2
55	7140	Shaft — Output	1	1
56	5701	"E" Ring 1/2	1	1
57	937158	Key — Woodruff Hi — Pro	1	1
58	6246	Chain — Roller 24 Pitches	1	1
59	6247	Chain — Roller 20 Pitches	1	1
60	910808-4	Bolt Round Head #10-32 x 5/8	1	1
61	910946-4	Bolt Round Head #10-32 x 2 3/4	2	2
62	5924	Nut Hex #10-32 Elastic Stop	11	11
63	7905	Collar — Shift	1	1
64	933190	Pin — Spirol 3/16 x 1 1/4	2	2
65	6252	Housing — Drive R.H.	1	1
66	6253	Housing — Drive L.H.	1	1
67	6254	Bearing	4	4
68	100279	Sprocket 17 Teeth	1	1
69	100278	Sprocket 8 Teeth	1	1
70	6261	Guide — Upper	1	1
71	6262	Guide — Lower	1	1
72	7864	Chain Roller 80 Pitches	1	1
73	6318	Screw #10-32 x 1 1/4 Fill. Head	6	6
74	920007-4	Washer — Plain 1/4 SAE	2	2
75	910068-4	Bolt Fill. Head 1/4-20 x 2 3/4	1	1
76	6390	Spring	1	1
77	7163	Bracket — Lever Shift	1	1
78	7133	Pin — Groove	1	1
79	7164	Lever — Shift	1	1
80	7165	Spacer	1	1
81	920005-4	Washer — Plain	1	1
82	910810-4	Screw Pan Head #10-32 x 7/8	1	1
83	911403-4	Bolt Hex #10-32 x 5/8	1	1
84	7922	Clip — Cable	1	1
85	933505-4	Hairpin	2	2
86	8608	Handle — Lower	1	1
87	8609	Handle — Upper	1	1
88	5523	Pin — Clevis 1/4 Dia.	2	2
89	900042-4	Bolt 5/16-18 x 2 Carriage	2	2
90	8610	Knob	2	2
91	8613	Panel — Handle	1	1
92	926235-4	Screw #10-16 x 5/8 Self Tap	2	2
93	8611	Throttle Control	1	1
94	920155-4	Lockwasher 5/16 Internal Tooth	2	2
95	8612	Clutch Control	1	1
96	8614	Plate — Clutch Control	1	1
97	100326	Knob	2	2
98	7098	Clip — Cable	2	2
99	910807-4	Screw #10-24 x 3/8 Pan Head	4	4
100	920120-4	Lockwasher #10 External Tooth	4	4
101	7657	Shroud — Engine	1	1
102	911793-4	Screw 1/4-28 x 1/2 Pan Head	4	4
103	8622	Plug — Cover	0	1
104	7674	Clip	0	1
105	8615	Hook	2	2
106	8616	Spring	2	2
107	3680	Cap	2	2
108	6227	Clip — Double	0	2
109	8617	Decal — Handle Panel	1	1
110	8618	Decal — Decoration — Panel	1	1
111	8619	Decal — Caution	1	1
112	9941	Clip — Battery Pack	0	1
113	5699	Decal — ASA Triangle	1	1
114	8621	Decal — Electric Start	0	1
115	6314	Decal — Chain Adjustment	1	1
116	8606	Decal — Self Propelled	1	0
117	8875	Decal — 22"	1	1
118	8624	Stud — Leaf Mulcher	2	2
119	8625	Plate	1	1
120	915584-4	Nut — Wing 1/4-20	2	2
121	8983	Bag	1	1
122	9183	Frame	1	1
123	8989	Hoop	1	1
124	8598	Belcrank Rear	1	1
125	9989	Battery — Power Pack	0	1
126	9956	Battery Charge	0	1
127	9957	Starter Cable	0	1
128	9958	Cap — Rope Pulley	0	1
129	9959	Key	0	2
130	9960	Bracket	0	1
131	910775-4	Screw Pan Head #8-32 x 3/8	0	4
132	100260	Decal — Caution	1	1

FIGURE 171   Wheel Horse Rotary Mower Models 3-1641, 3-1651, 3-1741, 3-1751, Nomenclature List

CATCHER ASSEMBLY

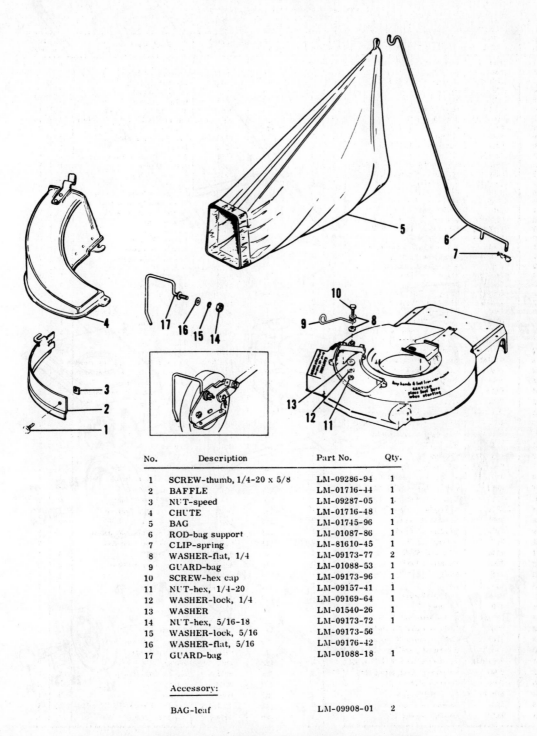

No.	Description	Part No.	Qty.
1	SCREW-thumb, 1/4-20 x 5/8	LM-09286-94	1
2	BAFFLE	LM-01716-44	1
3	NUT-speed	LM-09287-05	1
4	CHUTE	LM-01716-48	1
5	BAG	LM-01745-96	1
6	ROD-bag support	LM-01087-86	1
7	CLIP-spring	LM-81610-45	1
8	WASHER-flat, 1/4	LM-09173-77	2
9	GUARD-bag	LM-01088-53	1
10	SCREW-hex cap	LM-09173-96	1
11	NUT-hex, 1/4-20	LM-09157-41	1
12	WASHER-lock, 1/4	LM-09169-64	1
13	WASHER	LM-01540-26	1
14	NUT-hex, 5/16-18	LM-09173-72	1
15	WASHER-lock, 5/16	LM-09173-56	
16	WASHER-flat, 5/16	LM-09176-42	
17	GUARD-bag	LM-01088-18	1

Accessory:

BAG-leaf   LM-09908-01   2

Order parts from Homelite, Portchester, N.Y. 10573

FIGURE 172   Homelite Mower Models M-19, M-19E, Exploded Parts View and Nomenclature

PUSH MODEL – 19"

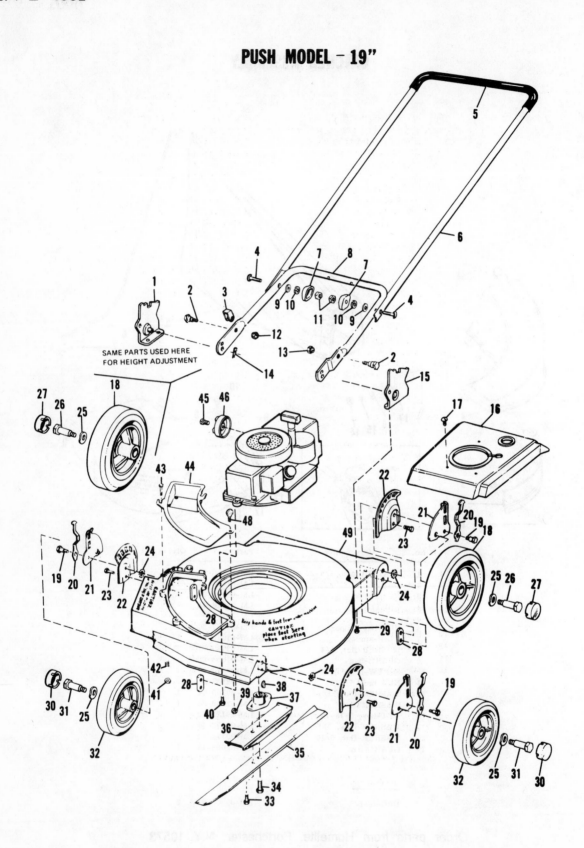

SAME PARTS USED HERE
FOR HEIGHT ADJUSTMENT

FIGURE 173   Homelite Push Model 19"

PUSH MODEL-19"

No.	Description	Part No.	M-19	M-19E	Qty.
1	BRACKET-(right hand)	LM-01718-18	x	x	1
2	BOLT-handle latch	LM-01088-34	x	x	2
3	LATCH-handle	LM-01087-27	x	x	1
4	BOLT-curved head	LM-01087-78	x	x	2
5	TUBING-handle	LM-01087-93	x	x	1
6	HANDLE-upper	LM-01088-47	x		1
	HANDLE-upper	LM-01721-65		x	1
7	NUT-wing	LM-01088-48	x	x	2
8	HANDLE-lower	LM-01717-80	x	x	1
9	WASHER-flat, 1/4	LM-09173-77	x	x	2
10	WASHER-lock, 5/16	LM-09173-56	x	x	2
11	NUT-hex, 5/16-18	LM-09286-98	x	x	2
12	NUT-hex, 5/16-18	LM-09233-62	x	x	1
13	NUT-handle latch	LM-01088-35	x	x	1
14	CLIP-spring	LM-81610-45	x	x	2
15	BRACKET-(left hand)	LM-01718-17	x	x	1
16	SHROUD	LM-01746-56	x	x	1
17	SCREW-rd. hd., 8 x 1/2	LM-09286-90	x	x	3
18	WHEEL & TIRE-8" (rear)	LM-01701-54	x	x	2
19	BOLT-hex shoulder	LM-01087-11	x	x	4
20	SPRING-height adjust	LM-01086-41	x	x	4
21	ADJUSTER-height	LM-01086-39	x	x	4
22	PLATE-adjusting	LM-01086-40	x	x	4
23	SCREW-hex cap, 3/8-16 x 5/8	LM-09182-28	x	x	4
24	WASHER-lock, 3/8	LM-09204-27	x	x	4
25	WASHER-spring	LM-01701-57	x	x	4
26	BOLT-wheel (rear)	LM-01701-35	x	x	2
27	CAP-hub (rear)	LM-01701-56	x	x	2
28	NUT-twin pilot	LM-01086-97	x	x	4
29	SCREW-hex, 1/4-20 x 3/8	LM-09274-28	x	x	4
30	CAP-hub (front)	LM-01701-55	x	x	2
31	BOLT-wheel (front)	LM-01087-64	x	x	2
32	WHEEL-6" (front)	LM-01701-53	x	x	2
33	SCREW-hex cap, 5/16-18 x 1	LM-09286-97	x	x	2
34	SCREW-hex cap, 3/8-24 x 1-1/2	LM-09221-28	x	x	1
35	BLADE-19"	LM-01088-50	x	x	1
36	STIFFENER-blade	LM-01087-04	x	x	1
37	ADAPTER-blade	LM-01087-16	x	x	1
38	KEY-Woodruff, #6	LM-09102-24	x	x	1
39	SPEED NUT-special	LM-01086-99	x	x	1
40	BOLT-engine mounting, 3/8-16 x 1	LM-09286-96	x	x	3
41	NUT-hex, 1/4-20	LM-09233-58	x	x	2
42	SPRING	LM-01716-51	x	x	2
43	SCREW-special	LM-01724-29	x	x	2
44	PLATE-cover	LM-01716-45	x	x	1
45	SCREW-hex, 8 x 1/2	LM-09280-06	x	x	1
46	BAFFLE-muffler	LM-01702-74	x	x	1
47	ENGINE-3-1/2 H. P.		x		1
	ENGINE-3-1/2 H. P.			x	1
48	SCREW-thumb, 1/4-20 x 5/8	LM-09286-94	x	x	1
49	HOUSING-19"	LM-01088-31	x	x	1

ALWAYS FURNISH MODEL AND SERIAL NUMBER WHEN ORDERING PARTS.

FIGURE 174   Homelite Push Model 19, 19E, Parts List

ELECTRIC START – 19"

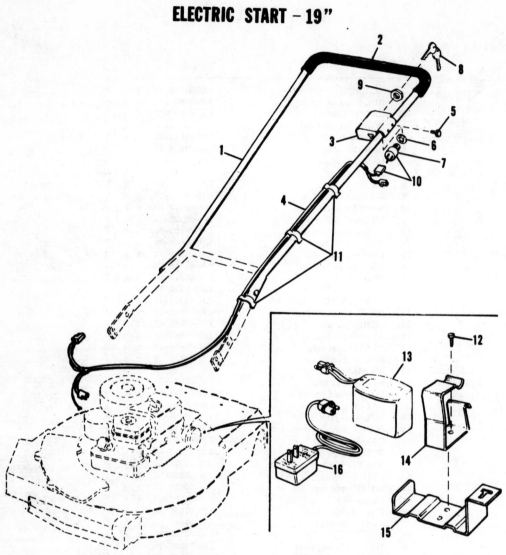

No.	Description	Part No.	Qty.
1	HANDLE-upper	LM-01721-65	1
2	TUBING-handle	LM-01087-93	1
3	BRACKET-switch mounting	LM-01714-77	1
4	HARNESS	LM-01714-75	1
5	SCREW-hex, 1/4-20 x 3/8	LM-09274-28	2
6	WASHER-lock, 9/16	LM-09286-36	1
7	SWITCH-starter	LM-01222-01	1
	Includes:		
8	KEY & RING	LM-01222-03	1
9	NUT-hex, special	LM-01222-35	1
10	SWITCH & HARNESS	LM-01718-02	1
11	CLAMP-handle cable	LM-01088-12	3
12	SCREW-pan, 10-24 x 3/8	LM-09276-63	2
13	BATTERY	LM-01724-15	1
14	CLIP-spring, battery	LM-01702-98	1
15	SUPPORT-battery	LM-01714-76	1
16	CHARGER-battery	LM-01724-31	1
	FILM-ignition (not shown)	LM-01718-00	1

ALWAYS FURNISH MODEL AND SERIAL NUMBER WHEN ORDERING PARTS.

FIGURE 175   Homelite Electric Start Model 19"

Part No.	Description	Qty.	Part No.	Description	Qty.
LM-01086-39	Adjuster-front height	4	LM-01749-14	Film-Homelite	1
LM-01086-40	Plate-adjusting	4	LM-01749-15	Film-M19E	2
LM-01086-41	Spring-adjuster	4	LM-01718-00	Film-ignition	1
LM-01087-04	Stiffener-blade	1	LM-01749-21	Film-stripe	1
LM-01087-16	Adapter-blade	1	LM-09102-24	Key-Woodruff #6	1
LM-01087-27	Latch-handle	1	LM-09908-01	Bag-leaf	2
LM-01087-79	Film-instruction	1		Plate-I. D.	1
LM-01087-86	Rod-bag support	1	LM-81610-45	Clip-spring	2
LM-01087-93	Tubing-handle	1	FASTENING PARTS		
LM-01088-12	Clamp-cable handle	3	LM-01086-97	Nut-pilot twin	4
LM-01088-18	Guard-bag	1	LM-01086-99	Speed Nut	1
LM-01088-31	Housing-19"	1	LM-01087-11	Bolt-shoulder	4
LM-01088-47	Handle-upper	1	LM-01087-64	Bolt-front wheel	2
	Engine	1	LM-01087-78	Bolt-curved head	2
LM-01088-50	Blade-19"	1	LM-01088-34	Bolt-latch handle	2
LM-01088-53	Guard-bag	1	LM-01088-35	Nut-latch handle	1
LM-01222-01	Switch-starter	1	LM-01088-48	Nut-wing	2
LM-01222-03	Key & Ring	1	LM-01222-35	Nut-hex	1
LM-01701-53	Wheel-front	2	LM-01540-26	Washer	1
LM-01701-54	Wheel & Tire-rear	2	LM-01701-35	Bolt-rear wheel	2
LM-01701-55	Cap-hub	2	LM-01701-57	Washer-spring	4
LM-01701-56	Cap-hub	2	LM-01724-29	Screw-special	2
LM-01702-74	Baffle-muffler	1	LM-09157-41	Nut-1/4-20	1
LM-01702-98	Clip-spring battery	1	LM-09169-64	Washer-1/4	1
LM-01714-75	Harness	1	LM-09173-56	Washer-5/16	3
LM-01714-76	Support-battery	1	LM-09173-72	Nut-5/16-18	1
LM-01714-77	Bracket-switch	1	LM-09173-77	Washer-1/4	4
LM-01716-44	Baffle	1	LM-09173-96	1/4-20 x 1	1
LM-01716-45	Plate-cover	1	LM-09176-42	Washer-5/16	1
LM-01716-48	Chute	1	LM-09182-28	3/8-16 x 5/8	4
LM-01716-51	Spring	2	LM-09204-27	Washer-3/8	4
LM-01717-80	Handle-lower	1	LM-09221-28	3/8-24 x 1-1/2	1
	Engine	1	LM-09233-58	Nut-1/4-20	2
LM-01718-02	Switch & Harness	1	LM-09233-62	Nut-5/16-18	1
LM-01718-17	Bracket-left hand	1	LM-09274-28	1/4-20 x 3/8	4
LM-01718-18	Bracket-right hand	1	LM-09276-63	10-24 x 3/8	2
LM-01721-65	Handle-upper	1	LM-09280-06	8 x 1/2	1
LM-01724-15	Battery	1	LM-09286-36	Washer-9/16	1
LM-01724-31	Charger-battery	1	LM-09286-90	8 x 1/2	3
LM-01729-80	Decal-safety	1	LM-09286-94	1/4-20 x 5/8	1
LM-01745-96	Bag	1	LM-09286-96	Bolt-3/8-16 x 1	3
LM-01746-20	Film-M19	2	LM-09286-97	Bolt-5/16-18 x 1	2
LM-01746-56	Shroud	1	LM-09286-98	Nut-5/16-18	2
			LM-09287-05	Nut-speed	1

FIGURE 176   Homelite Electric Start 19", Parts List

ELECTRIC START

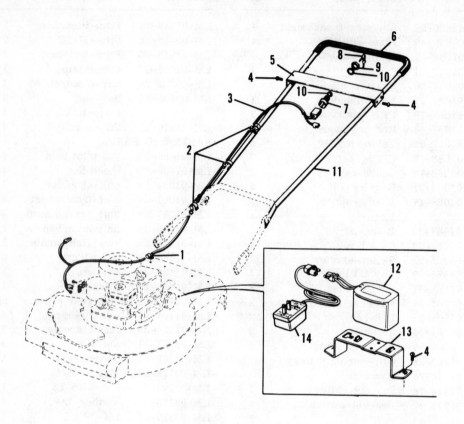

No.	Description	Part No.	M-21E	M-21SE	Qty.
1	CLAMP-cable	LM-01088-11	x	x	1
2	CLAMP-handle cable	LM-01088-12	x	x	3
3	HARNESS	LM-01703-00	x	x	1
4	SCREW-hex, 1/4-20 x 3/8	LM-09274-28	x	x	6
5	PANEL-switch mounting	LM-01703-02	x	x	1
6	TUBING-handle	LM-01087-93	x	x	1
7	SWITCH-starter	LM-01222-01	x	x	1
	Includes:				
8	KEY & RING	LM-01222-03	x	x	1
9	NUT-hex, special	LM-01222-35	x	x	1
10	WASHER-lock, 9/16	LM-09286-36	x		1
	WASHER-lock, 9/16	LM-09286-36		x	2
11	HANDLE-upper	LM-01086-55	x	x	1
12	BATTERY	LM-01088-52	x	x	1
13	SUPPORT-battery	LM-01700-14	x	x	1
14	CHARGER-battery	LM-01700-46	x	x	1

| ALWAYS FURNISH MODEL AND SERIAL NUMBER WHEN ORDERING PARTS.

FIGURE 177 Homelite Mower Models M-21, M-21E, M-21S, M-21SE, Exploded Parts Views and Nomenclature

PUSH MOWER –21

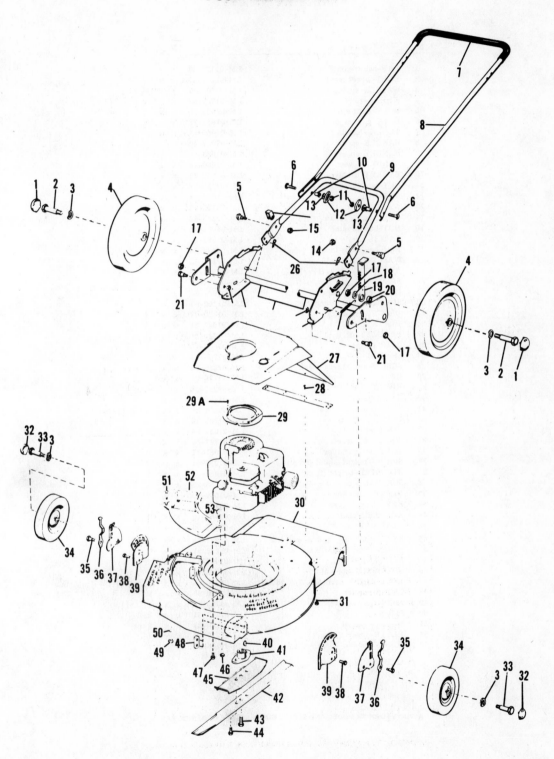

ALWAYS FURNISH MODEL AND SERIAL NUMBER WHEN ORDERING PARTS.

FIGURE 178   Homelite Push Model 21"

PUSH MOWER –21

No.	Description	Part No.	M-21	M-21E	Qty.
1	CAP-hub (rear)	LM-01701-56	x	x	2
2	BOLT-wheel (rear)	LM-01701-35	x	x	2
3	WASHER-spring	LM-01701-57	x	x	4
4	WHEEL-10" (rear)	LM-01721-09	x	x	2
5	BOLT-handle latch	LM-01088-34	x	x	2
6	BOLT-curved head	LM-01087-78	x	x	2
7	TUBING-handle	LM-01087-93	x	x	1
8	HANDLE-upper	LM-01087-45	x		1
	HANDLE-upper	LM-01086-55		x	1
9	HANDLE-lower	LM-01086-56	x	x	1
10	EYELET-handle	LM-01088-21	x	x	2
11	NUT-hex, 5/16-18	LM-09286-98	x	x	2
12	NUT-wing	LM-01088-48	x	x	2
13	WASHER-flat, 1/4	LM-09173-77	x	x	2
14	NUT-hex, 5/16-18	LM-09233-62	x	x	1
15	NUT-handle latch	LM-01088-35	x	x	1
16	LATCH-handle	LM-01087-27	x	x	1
17	NUT-hex, 3/8-16	LM-09234-28	x	x	6
18	WASHER-flat, 5/16	LM-09176-42	x	x	1
19	ADJUSTER-height (rear)	LM-01086-38	x	x	1
20	SPACER	LM-01051-84	x	x	1
21	BOLT-hex shoulder	LM-01087-66	x	x	2
22	SPRING-height adjust (rear)	LM-01702-64	x	x	1
23	BRACKET-(L.H.)	LM-01720-09	x	x	1
24	PLATE & TUBE	LM-01087-21	x	x	1
25	BRACKET-(R.H.)	LM-01720-11	x	x	1
26	CLIP-spring	LM-81610-45	x	x	2
27	SHROUD	LM-01746-54	x	x	1
28	SCREW-hex, 12-24 x 1/2	LM-09286-95	x	x	4
29	RING	LM-01702-97	x	x	1
29A	RIVETS	LM-09287-08	x	x	4
30	HOUSING-21"	LM-01700-13	x	x	1
31	SCREW-hex, 1/4-20 x 3/8	LM-09274-28	x	x	4
32	CAP-hub (front)	LM-01701-55	x	x	2
33	BOLT-wheel (front)	LM-01087-64	x	x	2
34	WHEEL-6" (front)	LM-01701-53	x	x	2
35	BOLT-hex shoulder	LM-01087-11	x	x	2
36	SPRING-height adjust (front)	LM-01086-41	x	x	2
37	ADJUSTER-height (front)	LM-01086-39	x	x	2
38	SCREW-hex cap, 3/8-16 x 5/8	LM-09182-28	x	x	2
39	PLATE-adjusting	LM-01086-40	x	x	2
40	KEY-Woodruff, #6	LM-09102-24	x	x	1
41	ADAPTER-blade	LM-01087-16	x	x	1
42	BLADE-21"	LM-01087-00	x	x	1
43	SCREW-hex cap, 3/8-24 x 1-1/2	LM-09221-28	x	x	1
44	SCREW-hex, 5/16-18 x 1	LM-09286-97	x	x	2
45	STIFFENER-blade	LM-01087-04	x	x	1
46	SPEED NUT-special	LM-01086-99	x	x	1
47	BOLT-engine mounting, 3/8-16 x 1	LM-09286-96	x	x	3
48	NUT-twin pilot	LM-01086-97	x	x	2
49	NUT-hex, 1/4-20	LM-09233-58	x	x	2
50	SPRING	LM-01716-51	x	x	2
51	SCREW-special	LM-01724-29	x	x	2
52	PLATE-cover	LM-01716-45	x	x	1
53	SCREW-thumb, 1/4-20 x 5/8	LM-09286-94	x	x	1

ALWAYS FURNISH MODEL AND SERIAL NUMBER WHEN ORDERING PARTS.

FIGURE 179   Homelite Push Model 21", Parts List

SELF PROPELLED MOWER – 21"
& ELECTRIC START

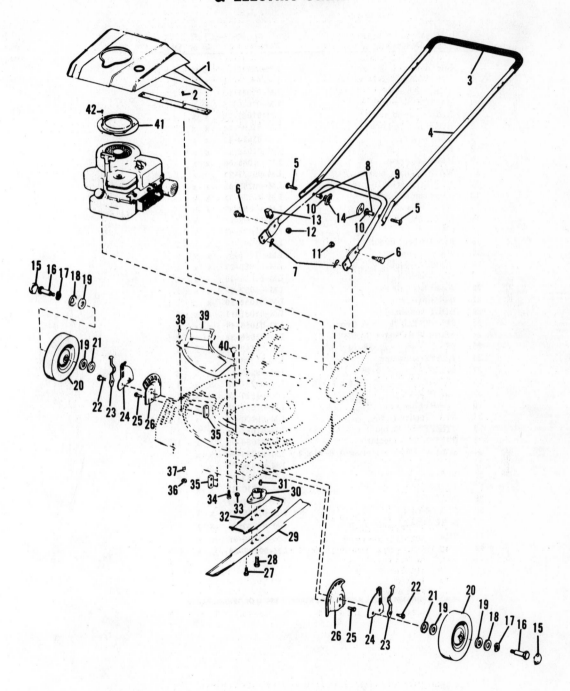

ALWAYS FURNISH MODEL AND SERIAL NUMBER WHEN ORDERING PARTS.

FIGURE 180   Homelite 21" Self Propelled and Electric Start

SELF PROPELLED MOWER −21"
& ELECTRIC START

No.	Description	Part No.	M-21S	M-21SE	Qty.
1	SHROUD	LM-01746-54	x	x	1
2	SCREW-hex, 12-24 x 1/2	LM-09286-95	x	x	4
3	TUBING-handle	LM-01087-93	x	x	1
4	HANDLE-upper	LM-01086-55	x	x	1
5	BOLT-curved head	LM-01087-78	x	x	2
6	BOLT-handle latch	LM-01088-34	x	x	2
7	CLIP-spring	LM-81610-45	x	x	2
8	EYELET-handle	LM-01088-21	x	x	2
9	HANDLE-lower	LM-01086-56	x	x	1
10	WASHER-flat, 1/4	LM-09173-77	x	x	2
11	NUT-hex, 5/16-18	LM-09233-62	x	x	1
12	NUT-handle latch	LM-01088-35	x	x	1
13	LATCH-handle	LM-01087-27	x	x	1
14	NUT-wing	LM-01088-48	x	x	2
15	CAP-hub	LM-01087-83	x	x	2
16	BOLT-wheel (front)	LM-01087-64	x	x	2
17	WASHER-spring	LM-01701-57	x	x	2
18	HOUSING-seal	LM-01088-27	x	x	2
19	SEAL-wheel	LM-01088-19	x	x	4
20	WHEEL-6" (front)	LM-01087-02	x	x	2
21	HOUSING-seal	LM-01088-20	x	x	2
22	BOLT-hex shoulder	LM-01087-11	x	x	2
23	SPRING-height adjust (front)	LM-01086-41	x	x	2
24	ADJUSTER-height (front)	LM-01086-39	x	x	2
25	SCREW-hex cap, 3/8-16 x 5/8	LM-09182-28	x	x	2
26	PLATE-adjusting	LM-01086-40	x	x	2
27	SCREW-hex, 5/16-18 x 1	LM-09286-97	x	x	2
28	SCREW-hex cap, 3/8-24 x 1-1/2	LM-09221-28	x	x	1
29	BLADE-21"	LM-01087-00	x	x	1
30	ADAPTER-blade	LM-01087-16	x	x	1
31	KEY-Woodruff, #6	LM-09102-24	x	x	1
32	STIFFENER-blade	LM-01087-04	x	x	1
33	SPEED NUT-special	LM-01086-99	x	x	1
34	SCREW-engine mounting, 3/8-16 x 1	LM-09286-96	x	x	3
35	NUT-twin pilot	LM-01086-97	x	x	2
36	NUT-hex, 1/4-20	LM-09233-58	x	x	2
37	SPRING	LM-01716-51	x	x	2
38	SCREW-special	LM-01724-29	x	x	2
39	PLATE-cover	LM-01716-45	x	x	1
40	SCREW-thumb, 1/4-20 x 5/8	LM-09286-94	x	x	1
41	RING	LM-01702-97	x	x	1
42	RIVETS	LM-09287-08	x	x	4

ALWAYS FURNISH MODEL AND SERIAL NUMBER WHEN ORDERING PARTS.

FIGURE 181   Homelite 21" Self Propelled and Electric Start, Parts List

SELF PROPELLED MOWER – 21"

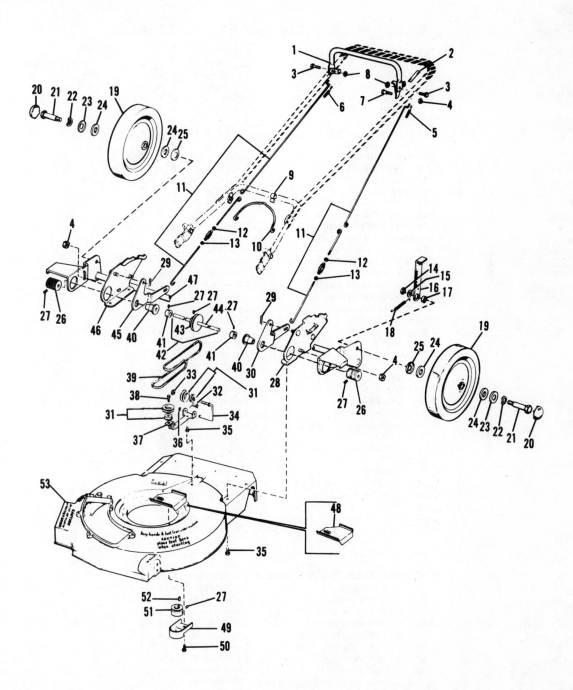

ALWAYS FURNISH MODEL AND SERIAL NUMBER WHEN ORDERING PARTS.

FIGURE 182   Homelite 21" Self Propelled

SELF PROPELLED MOWER – 21"

No.	Description	Part No.	M-21S	M-21SE	Qty.
1	HANDLE-engaging	LM-01087-54	x	x	1
2	FILM-handle	LM-01088-00	x	x	1
3	BOLT-handle shoulder	LM-01087-34	x	x	2
4	NUT-flange, 5/16-18	LM-09257-63	x	x	1
5	SPRING-rod return (L.H.)	LM-01087-91	x	x	1
6	SPRING-rod return (R.H.)	LM-01087-76	x	x	1
7	STOP-handle	LM-01087-57	x	x	1
8	NUT-hex, 5/16-18	LM-09286-98	x	x	4
9	CLAMP-cable handle	LM-01088-12	x	x	1
10	SPRING-rod	LM-01088-38	x	x	1
11	ROD-engaging	LM-01088-24	x	x	2
12	NUT-hex, 10-24 (L.H. thd.)	LM-09287-04	x	x	2
13	NUT-hex, 10-24	LM-09166-21	x	x	2
14	NUT-hex, 3/8-16	LM-09234-28	x	x	2
15	WASHER-flat, 5/16	LM-09176-42	x	x	1
16	ADJUSTER-height (rear)	LM-01086-38	x	x	1
17	SPACER-height adjust	LM-01051-84	x	x	1
18	SPRING-height adjust (rear)	LM-01702-64	x	x	1
19	WHEEL-10" (rear)	LM-01087-01	x	x	2
20	CAP-hub	LM-01087-83	x	x	2
21	BOLT-wheel (rear)	LM-01087-65	x	x	2
22	WASHER-spring	LM-01701-57	x	x	2
23	HOUSING-seal	LM-01088-20	x	x	2
24	SEAL-wheel	LM-01088-19	x	x	4
25	HOUSING-seal	LM-01088-27	x	x	2
26	DRIVER-wheel	LM-01086-64	x	x	2
27	SCREW-set, 5/16-18 x 1/4	LM-09287-34	x	x	6
28	BRACKET-(L.H.)	LM-01718-91	x	x	1
29	PIN-cotter, 3/32 x 5/8	LM-09220-00	x	x	2
30	BELLCRANK-(L.H.)	LM-01087-36	x	x	1
31	PULLEY-idler	LM-01087-30	x	x	2
32	SPRING-idler torsion	LM-01087-29	x	x	1
33	NUT-flange, 1/4-20	LM-09267-06	x	x	1
34	U-BOLT-idler	LM-01086-93	x	x	1
35	SCREW-hex, 1/4-20 x 3/8	LM-09274-28	x	x	6
36	RING-"E" Type	LM-01086-98	x	x	1
37	BRACKET-idler support	LM-01086-94	x	x	1
38	SCREW-hex, 1/4-20 x 3/4	LM-09255-92	x	x	1
39	BELT-drive	LM-01087-32	x	x	1
40	BEARING & RETAINER	LM-01086-67	x	x	2
41	COLLAR-set	LM-83010-42	x	x	2
42	SHAFT & PULLEY	LM-01088-23	x	x	1
	Includes:				
43	PULLEY-drive	LM-01086-88	x	x	1
44	SHAFT-wheel driver	LM-01086-66	x	x	1
45	BELLCRANK-(R.H.)	LM-01086-30	x	x	1
46	BRACKET-(R.H.)	LM-01718-93	x	x	1
47	PLATE & TUBE	LM-01086-36	x	x	1
48	DEFLECTOR-oil	LM-01088-41	x	x	1
49	COVER-drive (lower)	LM-01087-03	x	x	1
50	SCREW-hex, 12-24 x 1/2	LM-09286-95	x	x	1
51	PULLEY-engine	LM-01086-89	x	x	1
52	KEY-Woodruff, #5	LM-09051-21	x	x	1
53	HOUSING-21"	LM-01700-13	x	x	1
	PLATE-identification (not shown)	LM-16070-48	x	x	1

ALWAYS FURNISH MODEL AND SERIAL NUMBER WHEN ORDERING PARTS.

FIGURE 183 Homelite 21" Self Propelled, Parts List

CATCHER ASSEMBLY

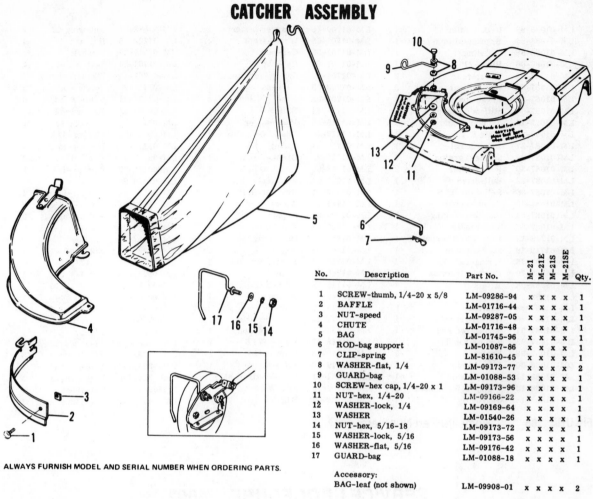

No.	Description	Part No.	M-21	M-21E	M-21S	M-21SE	Qty.
1	SCREW-thumb, 1/4-20 x 5/8	LM-09286-94	x	x	x	x	1
2	BAFFLE	LM-01716-44	x	x	x	x	1
3	NUT-speed	LM-09287-05	x	x	x	x	1
4	CHUTE	LM-01716-48	x	x	x	x	1
5	BAG	LM-01745-96	x	x	x	x	1
6	ROD-bag support	LM-01087-86	x	x	x	x	1
7	CLIP-spring	LM-81610-45	x	x	x	x	1
8	WASHER-flat, 1/4	LM-09173-77	x	x	x	x	2
9	GUARD-bag	LM-01088-53	x	x	x	x	1
10	SCREW-hex cap, 1/4-20 x 1	LM-09173-96	x	x	x	x	1
11	NUT-hex, 1/4-20	LM-09166-22	x	x	x	x	1
12	WASHER-lock, 1/4	LM-09169-64	x	x	x	x	1
13	WASHER	LM-01540-26	x	x	x	x	1
14	NUT-hex, 5/16-18	LM-09173-72	x	x	x	x	1
15	WASHER-lock, 5/16	LM-09173-56	x	x	x	x	1
16	WASHER-flat, 5/16	LM-09176-42	x	x	x	x	1
17	GUARD-bag	LM-01088-18	x	x	x	x	1

Accessory:
BAG-leaf (not shown) LM-09908-01 x x x x 2

ALWAYS FURNISH MODEL AND SERIAL NUMBER WHEN ORDERING PARTS.

FIGURE 184   Homelite Catcher Assembly

Part No.	Description	Qty.	Part No.	Description	Qty.	Part No.	Description	Qty.
LM-01051-84	Spacer	1	LM-01088-19	Seal-wheel	4			
LM-01086-30	Bellcrank-R. H.	1	LM-01088-20	Housing-seal	2	LM-81610-45	Clip-spring	2
LM-01086-36	Plate & Tube	1	LM-01088-21	Eyelet-handle	2	LM-83010-42	Collar-set	2
LM-01086-38	Adjuster-rear	1	LM-01088-23	Shaft & Pulley	1	FASTENING PARTS		
LM-01086-39	Adjuster-front	2	LM-01088-24	Rod-engaging	2	LM-01086-97	Nut-pilot	2
LM-01086-40	Plate-adjusting	2	LM-01088-27	Housing-seal	2	LM-01086-99	Nut-speed	1
LM-01086-41	Spring-adjuster	2	LM-01088-38	Spring-rod	1	LM-01087-11	Bolt-hex	2
LM-01086-55	Handle-upper	1	LM-01088-41	Deflector-oil	1	LM-01087-34	Bolt-shoulder	2
LM-01086-56	Handle-lower	1	LM-01088-52	Battery	1	LM-01087-64	Bolt-front	2
LM-01086-64	Driver-wheel	2	LM-01088-53	Guard-bag	1	LM-01087-65	Bolt-rear	2
LM-01086-66	Shaft-wheel driver	1	LM-01222-01	Switch-starter	1	LM-01087-66	Bolt-shoulder	2
LM-01086-67	Bearing & Retainer	2	LM-01222-03	Key & Ring	1	LM-01087-78	Bolt-curved	2
LM-01086-88	Pulley-drive	1	LM-01700-13	Housing-21"	1	LM-01088-34	Bolt-latch	2
LM-01086-89	Pulley-engine	1	LM-01700-14	Support-battery	1	LM-01088-35	Nut-latch	1

LM-01086-93	U-Bolt Idler	1	LM-01700-46	Charger-battery	1	LM-01088-48	Nut-wing	2
LM-01086-94	Bracket-idler support	1	LM-01701-53	Wheel-front	2	LM-01222-35	Nut-hex	1
LM-01086-98	"E" Ring	1	LM-01701-55	Cap-hub	2	LM-01540-26	Washer	1
LM-01087-00	Blade-21"	1	LM-01701-56	Cap-hub	2	LM-01701-35	Bolt-wheel	2
LM-01087-01	Wheel-rear	2	LM-01702-64	Spring-adjuster	1	LM-01701-57	Washer-spring	4
LM-01087-02	Wheel-front	2	LM-01703-00	Harness	1	LM-01724-29	Screw-special	2
LM-01087-03	Cover-lower drive	1	LM-01703-02	Panel-switch	1	LM-09287-34	5/16-18 x 1/4	6
LM-01087-04	Stiffener-blade	1	LM-01716-44	Baffle	1	LM-09166-22	Nut-1/4-20	1
LM-01087-16	Adapter-blade	1	LM-01716-45	Plate-cover	1	LM-09166-21	Nut-10-24	2
LM-01087-21	Plate & Tube	1	LM-01716-48	Chute Assembly	1	LM-09169-64	Washer-1/4	1
LM-01087-27	Latch-handle	1	LM-01716-51	Spring	2	LM-09173-56	Washer-lock	1
LM-01087-29	Spring-Torsion	1	LM-01718-91	Bracket-L. H.	1	LM-09173-72	Nut-5/16-18	1
LM-01087-30	Pulley-idler	2	LM-01718-93	Bracket-R. H.	1	LM-09173-77	Washer-1/4	2
LM-01087-32	Belt-drive	1	LM-01720-09	Bracket-L. H.	1	LM-09173-96	1/4-20 x 1	1
LM-01087-36	Bellcrank-L. H.	1	LM-01720-11	Bracket-R. H.	1	LM-09176-42	Washer-5/16	1
LM-01087-45	Handle-upper	1	LM-01721-09	Wheel-rear	2	LM-09182-28	Screw-hex cap	2
LM-01087-54	Handle-engaging	1	LM-01724-94	Engine	1	LM-09220-00	Pin-cotter	2
LM-01087-57	Stop-handle	1	LM-01727-63	Film-latch	1	LM-09221-28	3/8-24 x 1-1/2	1
LM-01749-13	Film-throttle control	1	LM-01729-80	Decal-safety	1	LM-09233-58	Nut-lock	2
LM-01749-19	Film-caution	1	LM-01745-96	Bag	1	LM-09233-62	Nut-hex lock	1
LM-01087-76	Spring-rod	1	LM-01746-06	Film-Homelite	1	LM-09234-28	Nut-3/8-16	6
LM-01087-79	Film-instruction	1	LM-01746-07	Film-M21	2	LM-09260-76	Nut-1/4-20	1
LM-09287-08	Rivets	4	LM-01746-12	Film-mtg. panel	1	LM-09257-63	Nut-5/16-18	1
LM-01702-97	Ring	1	LM-01746-54	Shroud	1	LM-09255-92	1/4-20 x 3/4	1
LM-01087-83	Cap-hub	2	LM-01749-12	Film	2	LM-09274-28	1/4-20 x 3/8	6
LM-01087-86	Rod-bag support	1	LM-01749-13	Film-throttle	1	LM-09286-36	Washer-9/16	2
LM-01087-91	Spring-ret. rod	1	LM-01753-40	Film-ignition	1	LM-09286-94	1/4-20 x 5/8	1
LM-01087-93	Tubing-handle	1	LM-01749-17	Film-M21 E	2	LM-09286-95	12-24 x 1/2	4
LM-01087-98	Engine	1	LM-01749-18	Film-M21S	2	LM-09286-96	3/8-16 x 1	3
LM-01088-00	Film-handle	1	LM-01749-20	Film-stripe	1	LM-09286-97	5/16-18 x 1	2
LM-01088-11	Clamp-cable	1	LM-09051-21	Key-Woodruff #5	1	LM-09286-98	Nut-5/16-18	4
LM-01088-12	Clamp-cable handle	3	LM-09102-24	Key-Woodruff #6	1	LM-09287-04	Nut-10-24	2
LM-01088-18	Guard-bag	1	LM-09908-01	Bag-leaf	2	LM-09287-05	Nut-Tinnerman	1

FIGURE 185   Homelite Part Numbers

SERVICE PROCEDURE   4002

Rotary Mower Blade Removal, Sharpening and Balance

Most lawn mower manufacturers stress the importance of keeping the rotary mower blade sharp and balanced. When the blade is sharp the grass is cleanly cut by the sicklelike action of the blade. As the blade dulls and is nicked from striking lawn debris the sickle action is lost and the blades of grass are literally torn away. The turf will reflect this since the blades of grass will have ragged rough ends which quickly brown. As a result of the ragged ends, the grass blades will lose more of the necessary plant nutrients than if cleanly cut, and healing of the rough ends is delayed much longer. Frequent examination and sharpening of the rotary blade is recommended for ease in cutting and if the lawn is to have that well-groomed look.

---- **CAUTION** ----

BEFORE ANY WORK OR INSPECTION OF THE ROTARY BLADE, BE ABSOLUTELY CERTAIN TO REMOVE THE IGNITION CABLE FROM THE SPARK PLUG.

---- **CAUTION** ----

REMOVAL OF BLADE

To examine and remove the blade, carefully place the machine on its side. Be sure that the oil fill port faces up. Be careful not to dump oil from port, nor fuel from the fuel tank. Refer to Figs. 186, 187 and 188 for typical rotary blade arrangements. Proceed to remove the blade as follows:

- Use an exact fitting socket wrench or box wrench to remove the blade fastener. Hold the blade firmly (wrapping a cloth over the sharp ends or wearing a leather glove prevents possible cuts or scratches) and turn the fastener counterclockwise.

- Note how the blade fastener, washer, adaptor washer, or anti-scalp disc, etc., are assembled as shown in the typical illustration of Fig. 186. Place all these parts in a can or pan for safekeeping.

- Examine the blade retainer for damage, e.g. being bent out of shape. Discard and use a new part (exact replacement) if this is the case. Carefully examine the key and keyway on the crankshaft. Look for very loose fit of key in keyway or heavy burrs which may keep the key from fitting properly. Replace with a new key if any of these problems are evident. Be certain to use an exact replacement.

SHARPENING

The blade can be sharpened by holding it in a table vise and using a sharp file to sharpen the cutting edges. File slowly, removing as little material as possible until the cutting edge is cleaned up of deep nicks or cuts. For blades with very deep nicks it is better to use a bench grinding wheel, if this is available. See Fig. 189.

Try to hold the same cutting angle on the blade as was there before you started sharpening. Be sure to clean the blade off before sharpening.

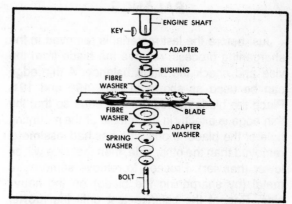

FIGURE 186 Typical Rotary Mower Blade Parts Removal

FIGURE 187 Rotary Mower Blade

FIGURE 188 Rotary Mower Blade

FIGURE 189 Mower Rotary Blade Bench Grinder Sharpener

BALANCE

Just before the last material is removed in the sharpening process, remove the blade from the vise and check the blade balance. A thin edge can be used as shown in Figs. 190 and 191. Place the blade across the thin edge so that the thin edge is exactly in the middle of the mounting hole of the blade. If one side has had less metal removed than the other side, then that side will be lower (heavier). Proceed to remove some more metal (by sharpening the blade) on the heavy side. Repeat the balance test until the mower blade remains level when tested on the knife edge.

Put a light film of machine oil on all the blade parts before reinstalling. Be careful to install the key, adaptor, bushing, etc., in the reverse order that these parts were removed. Hold the blade firmly and tighten the blade fastener securely.

Turn the machine upright and replace ignition cable on spark plug.

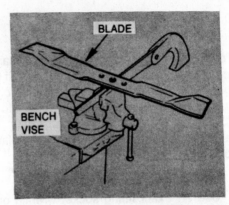

FIGURE 190   Balancing Sharpened Mower Blade

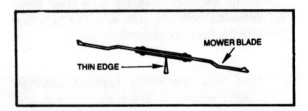

FIGURE 191   Thin Edge Used as Blade Balancer

SERVICE PROCEDURE   4003

Reel Mower Blade Balance and Sharpening

Reel mower blades when damaged by being bent out of the normal shape or upon loss of some of the cutting edge of the blade require reshaping and balancing. Rebalance of the reel is a complex task since the reel must be completely removed, repaired and then balanced on special test equipment. Usually this is an expensive procedure and the reel may not perform the same even then as before it was damaged. Therefore a bent, damaged reel is best discarded and replaced by a new reel.

SHARPENING REEL BLADES

a. The reel lawn mower blade requires the use of a sharpening machine of the type shown in Fig. 192. Machines of this type are generally not found in the home workshop. The service of a mower sharpening shop is, therefore, required.

b. However, there is one fairly simple and very economical method which can be undertaken by the lawn mower owner. If the blades are not badly in need of sharpening the reel can be sharpened by using a light coat of lapping or grinding compound placed on the bed knife. The reel blade is then adjusted such that the blades come into light contact with the bed knife. The reel is then turned *backwards by hand* and in so doing, the blade, and the bed knife are ground with the grinding compound.

NOTE:

To turn the reel backwards the drive chain, sprocket or pulley must first be disengaged or removed.

FIGURE 192   Reel Mower Sharpening Machine

SERVICE PROCEDURE   4004

Spark-plug Removal, Cleaning, Regapping, Replacement

Perhaps the most frequent part of the lawn mower engine which is subject to overhaul or replacement is the spark plug. The spark plug is very simple to service.

─────────**CAUTION**─────────

DO NOT WORK ON
A HOT ENGINE.
WAIT UNTIL ENGINE
FULLY COOLS.

─────────**CAUTION**─────────

Clean away all dirt, grease or grass clippings from the engine head where the spark plug is located. Determine the spark-plug size (from the manufacturer's instruction manual, or read the number and brand on the spark plug) and use a spark-plug socket wrench *to fit that specific size.* The plug socket wrench must fit snug but not tight. If the socket fits loosely, it is the wrong size socket. See Figs. 193, 194, 195 and 196 for illustrations of various spark-plug sockets and socket wrench ratchet handle.

SPARK-PLUG REMOVAL

a.   Remove the ignition cable from the spark plug. Slip the socket wrench over the spark plug. Be sure the socket goes down over the plug as far as it will go. Unscrew by turning the wrench counterclockwise.

b.   Screw out the spark plug. See Fig. 197. Examine the plug hole. If it appears to be very dirty, or the plug came out very hard—was tight unscrewing almost all the way—then the plug hole threads should be cleaned. Use a spark-plug hole thread and seat cleaning tool as shown in Fig. 198.

c.   Remove the spark plug from the socket and inspect closely. Refer to Fig. 13 which shows a cross-section of a typical spark plug and plug part nomenclature. Examine the spark plug and refer to Fault Indication **2100** for a description of the various types of plug fouling that can be experienced. Discard the plug as indicated per the descriptions of **2100** if these apply to the plug. See Table III for a listing of lawn mower spark-plug types and plug manufacturer conversion numbers. If the plug can be cleaned, proceed as follows.

FIGURE 196   Typical Spark-Plug Socket

FIGURE 193   Spark Plug Socket Wrench

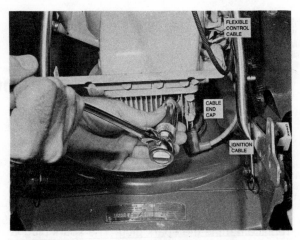

FIGURE 197   Unscrewing Spark Plug Using Socket Wrench

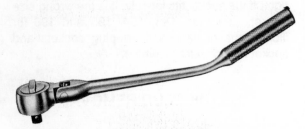

FIGURE 194   Socket Wrench Ratchet Handle

CLEANING

a. Hold the plug firmly and using a stiff wire brush, brush away the accumulations from the electrodes. Use a small can to hold some cleaning solvent and dip the gap end of the plug into the solvent. Continue to brush and wipe off the electrodes with a cloth. Use a wire or hairpin to get inside or around the insulator to scrape away any deposits. Brush the threads clean with the wire brush, or if available use a wire wheel as shown in Fig. 199. The plug should look like Fig. 200 when it is clean.

FIGURE 195   Typical Spark-Plug Socket

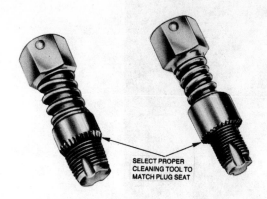

FIGURE 198   Spark-Plug   Hole   Thread
Cleaning Tools

b. Before gapping use a small flat distributor
point file to *very lightly* file/clean the center
electrode as shown in Fig. 201. A sharp
edge facilitates the propagation of the
spark.

TABLE III

POWER LAWNMOWER SPARK-PLUG CONVERSION CHART

AC Plug No. Plug Type	Autolite Type	Champion Type
LM 46	AU 7 PM	J-17 LM
CS 45	47 N	C J8
SR 45 LE	—	EH-10
G C46	A 71	—
M 45	A 7X	J 8 J
		J 11 J
C 86	BT 8	D–16
M 44 C	A 3 X	J–6J
M 47	A 9 X	J–12J
C 43 L	ATL 4	H–4
		H–8
45	A7	J–8
C 87	BT 9	D–21

GAPPING

a. Check the engine manufacturer's specification for proper spark-plug gap size. If this is
not known, ask the Authorized Factory
Service Dealer what the spark-plug gap is
for the engine. Use a spark-plug gap tool as
shown in Figs. 201 and 202 and measure
the gap as shown in Fig. 203. A round wire
gage should be used to check the true gap.
Figs. 204 and 205 illustrate the differences
in measurement which can occur using a
plain flat feeler gage and a round wire gage.

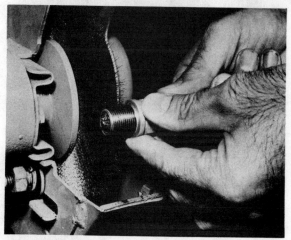

FIGURE 199   Cleaning Plug Threads Using
Wire Wheel

FIGURE 200   Cleaned Spark Plug

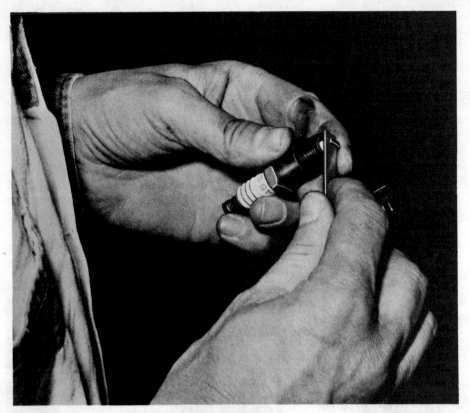

FIGURE 201   Light Filing of Electrodes

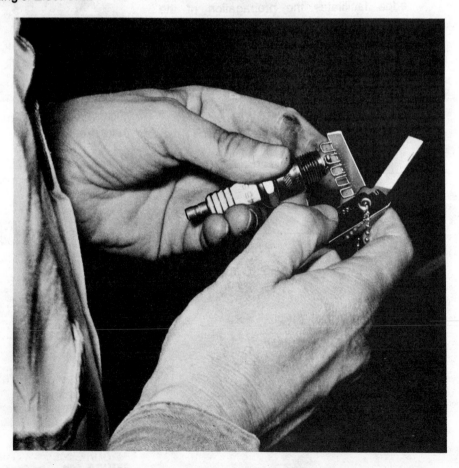

FIGURE 202   Typical Spark-Plug Gap Measuring Tool

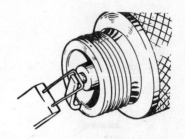

FIGURE 203 Gap Measurement Using Round Wire Gage

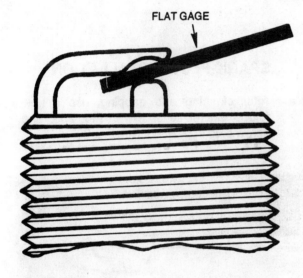

FLAT GAGE

FIGURE 204 Using Flat Gage to Measure Gap Results in Error in Reading

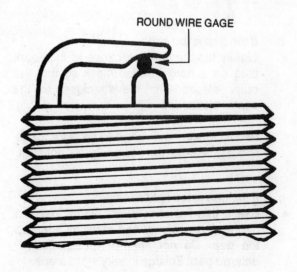

ROUND WIRE GAGE

FIGURE 205 Using Round Wire Gage to Measure Gap Results in Accurate Reading

┌─────── **CAUTION** ───────┐

NEVER ATTEMPT
TO ENLARGE OR MOVE
THE CENTER ELECTRODE.
THIS CAN CHIP
OR CRACK THE INSULATOR
AND RENDER THE PLUG USELESS.

└─────── **CAUTION** ───────┘

NOTE: NEW
GASKET IS NOT
COMPRESSED

FIGURE 206   Use of New Folded Spark-Plug Gasket

SPARK-PLUG INSTALLATION

a. After cleaning and gapping, use a new folded gasket as shown in Fig. 206, and screw plug in by hand. See Fig. 207.

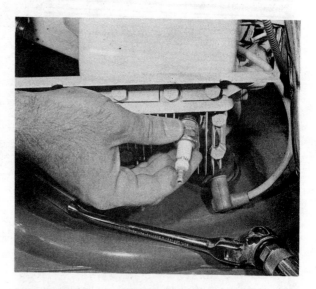

FIGURE 207   Screw Spark Plug in by Hand to Insure Proper Thread Start

b. *If the gap is too large:*
*   *Lightly* tap the outer electrode of the spark plug on a hard wooden block so that the outer electrode is moved closer to the center electrode.

*   Recheck with the gage. The wire gage should just enter the gap snug, *but not tight.*

c. *If the gap is too tight or small:*
*   Use a thin flat screwdriver blade; insert in gap and very slowly and carefully open up the gap. *Do not attempt to significantly enlarge gap.* Enlarge in very small steps.

*   Keep checking the gap with the gage, until the wire gage just enters. *Then stop opening* up the gap.

b. If a torque wrench is unavailable Table IV provides information, for cast iron and aluminum heads, on how the desired torque can be achieved.

c. Reconnect the ignition cable to the plug.

TABLE IV

SPARK PLUG INSTALLATION

Plug Thread	TORQUE WRENCH Cast Iron Heads	Aluminum Heads	NORMAL WRENCH Cast Iron or Aluminum Heads
10 mm.	12 lbs. ft.	10 lbs. ft.	3/8 to 1/2 turn
12 mm.	18 lbs. ft.	16 lbs. ft.	1/4 turn
14 mm.	25 lbs. ft.	22 lbs. ft	1/2 to 3/4 turn
14 mm. Taper Seat	15 lbs. ft.	15 lbs. ft.	———
18 mm.	35 lbs. ft.	25 lbs. ft.	1/2 to 3/4 turn
18 mm. Taper Seat	17 lbs. ft.	15 lbs. ft.	———
7/8"	40 lbs. ft.	30 lbs. ft.	1/2 to 3/4 turn

The above values apply when threads on spark plug and port are clean, the spark plug has been installed "finger tight," and (except tapered seats) a new folded gasket is used. Spark plugs with solid gaskets always require the use of a torque wrench.

SERVICE PROCEDURE   4005

Electric Battery, Wet Cell and Sealed Types

WET CELL BATTERIES

Lead-acid storage batteries are wet cell batteries. In these batteries the volume of electrolyte must be maintained to ensure a satisfactory power output capability. Perform the following maintenance services at regular intervals, and especially at the start and end of the mowing season.

a. Check electrolyte level. Level should be above the plates, or to the level marked on the battery cell caps. *DO NOT OVERFILL.* Overfill causes loss of electrolyte and can contribute to poor power output.

b. Keep battery top and terminals clean, free of corrosion. Use a solution of baking soda and water to wash away acid corrosion at terminals. Wash, and then rinse terminals with water.

─── CAUTION ───

BE CAREFUL NOT TO ALLOW SODA SOLUTION TO ENTER THE CELLS.

─── CAUTION ───

NOTE:

Be certain the battery is connected in accordance with the mower manufacturer's specifications. Most mower engines use a negative ground system in which the negative (−) terminal of the battery is connected in a common ground system with the metal portions of the engine. A heavy braided metal strap, or cable, connects this negative terminal to the engine. The positive (+) terminal is the "live" terminal and is connected to the mower starting switch and electric starter motor. The positive (+) terminal is also the larger of the two terminals, is usually stamped with a (+) and/or painted red on top. For those mower engines which use a positive (+) ground system the reverse of the negative system connection will be the case. That is, the positive (+) terminal has a heavy braided metal strap, or cable, connecting the (+) terminal to the engine. When disconnecting the battery in either type system, always remove the ground terminal connection first. See Fig. 208 for a sketch of the connection of a typical negative (−) ground system. The symbol (⏚) denotes the common ground of the engine and is physically that point where the ground cable (or braided metal strap) is mechanically attached to the engine.

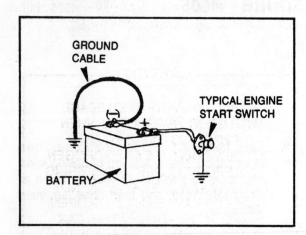

FIGURE 208   Typical Negative (−) Ground Connections

BATTERY TESTS

The following tests should be performed to determine the condition of the battery. Remove the battery from the mower.

- *VOLTAGE TEST:* Using a volt/ammeter as shown in Fig. 209, measure the voltage of each cell. For a 12 volt, 6 cell battery each cell contributes about 1.95 to 2.08 volts to the total battery voltage of 12 volts. If a difference of more than 0.05 volts between the highest and lowest cell voltage is obtained you may have internal cell damage. If the battery does not hold a charge on recharging it must be replaced. If the voltage is low but the high and low difference is less than .05 volts the battery will take a charge. See BATTERY CHARGING.

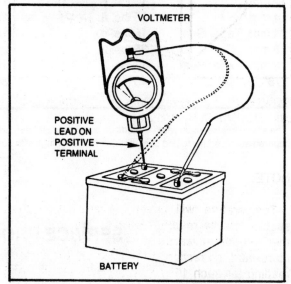

FIGURE 209   Use of Voltmeter in Battery Test

- *HYDROMETER TEST: Using a hydrometer, as shown in Fig. 210, measure the specific gravity of the electrolyte in each cell.*

  When the battery discharges and is not recharged, the amount of sulfuric acid in the electrolyte goes down and lead sulfate is formed as a result of chemical change in the battery. The lead sulfate is deposited on the battery plates, and the overall result is a lower specific gravity reading of the electrolyte. If the specific gravity reading in the cell falls below 1.240 the battery should be recharged. When fully charged the reading of the electrolyte specific gravity should be between 1.260 to 1.280.

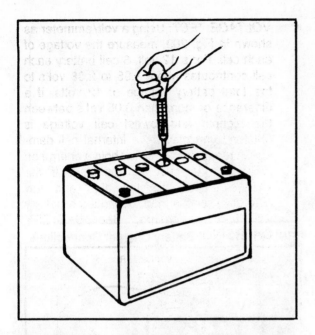

FIGURE 210   Measurement of Battery Specific Gravity with Hydrometer

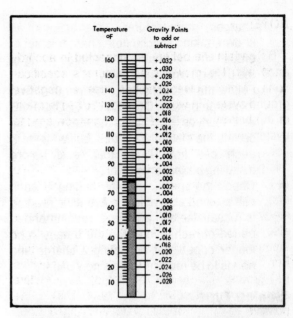

FIGURE 211   Specific Gravity Reading Correction Chart and Example of Use

Example:
Hydrometer Reading 1.255
Acid Temperature ...... 100 degrees Fahrenheit
Add Specific Gravity Correction .... .008
Corrected Specific Gravity is 1.263

A fully charged bettery has a specific gravity reading of 1.260 plus .015 minus .005 (all batteries for use in temperate climates).

NOTE:

Temperature will cause a variation in the specific gravity reading. Refer to Fig. 211 for a thermometer indication of the electrolyte temperature and correction for adding 0.004 to the reading for each 10° above 80° Fahrenheit, and subtracting 0.004 for each 10° below 80° Fahrenheit.

BATTERY CHARGING

When the battery is recharged the chemical reaction inside the battery is reversed wherein the lead sulfate deposits on the plates become disassociated and change back to lead, lead dioxide, and sulfuric acid. Thus the amount of sulfuric acid in the electrolyte is increased and this chemical process restores the battery to a condition where it can again supply electrical power. However it should be noted if there is internal damage to the plates or the walls of the cells, or if too much lead sulfate is deposited on the plates the battery may be permanently damaged and will need to be replaced with a new one. Referring to Fig. 212 proceed as follows to charge the battery.

---— **CAUTION** ---—

PROVIDE A LARGE OPEN SPACE WITH GOOD VENTILATION WHEN CHARGING BATTERY. AN EXPLOSIVE MIXTURE OF HYDROGEN GAS IS PRODUCED WHILE CHARGING BATTERY AT A HIGH RATE. DO NOT SMOKE OR HAVE SPARKS OR OPEN FLAME IN THE AREA AS THIS CAN CAUSE INTERNAL EXPLOSION IN BATTERY AND/OR FIRE.

---— **CAUTION** ---—

• Remove the battery cell caps, fill each cell (to indicated level) with distilled water.
• Hook up a battery charger, follow the

charger manufacturer's instructions as shown, taking special note charger is set at proper voltage setting (the voltage rating of the battery, usually 12 volts).

- Follow the instructions for the length of time to charge battery. DO NOT OVERCHARGE as this drives water out of the electrolyte. Be certain the electrolyte is at specified level in each cell and fill to that level before installing battery on mower.

- Check the specific gravity reading in each cell as cited above. In the event the reading is unsatisfactory after the recommended period of recharging then the battery is no longer capable of accepting a charge and needs to be replaced with a new battery.

Sealed Type

The sealed type battery differs from the wet cell lead-acid battery in that there is no liquid electrolyte volume to add to and maintain. This type is sealed and is self contained requiring no external materials. It provides electric power by chemical reactions as does the lead-acid type. Like the acid type this battery does require recharging in order to keep its electric power capability up to the point where it can drive the mower starter motor.

BATTERY TEST

Most sealed batteries, or power packages as they are called by a number of manufacturers, are not available with tester units for measuring the voltage capability of the battery. Such test units are special equipments due to unique cable connections on the package and are usually available at Authorized Service Dealers. If required the battery can be tested by the Service Dealer. However power packages do come equipped with a charger unit. By following a regular practice of plugging the battery charger unit into the recommended 115 volt, AC, single phase, 60 cycle outlet, after each use of the mower, a long efficient performance of the battery is ensured. Figs. 213 and 214 show some typical power packages and charger units.

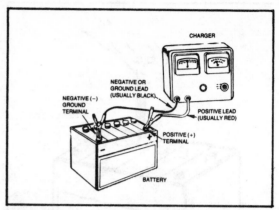

FIGURE 212   Battery Charger Connections

FIGURE 213   Typical Power Package

FIGURE 214   Typical Battery Power Pack Under Charge

NOTE:

Do not leave the power package plugged into the charger unit for an extended time period (over 48 hours). The charger should be disconnected during winter storage.

SERVICE PROCEDURE   4010

Carburetor:

Removal and Overhaul

The following procedures are written for application on many of the different carburetors which are in use on gasoline engine lawn mowers. Therefore the information given is general in nature. If the carburetor on your particular engine is not shown, the procedure will still apply and permit servicing of the carburetor.

Figs. 215 to 220 show a number of different manufacturers' designs of carburetors. Most carburetors follow one of two basic designs. One design uses a suction lift which draws up the fuel from a fuel tank. This type carburetor is mounted directly on top of the fuel tank. The other design is supplied fuel through a fuel line. The fuel supply may be gravity feed or it can be pumped to the carburetor by use of a fuel pump. Fig. 221 shows a cut away drawing of a suction lift type carburetor. Other figures which follow will show exploded line drawings of the individual parts and components of carburetors. Proceed to Step 1 for removal of the carburetor from the engine. Before proceeding farther the following caution should be strictly observed.

---CAUTION---

DO NOT PERFORM ANY CARBURETOR WORK OR REPAIR ON A HOT ENGINE. WAIT UNTIL ENGINE FULLY COOLS OFF. DISCONNECT THE IGNITION CABLE THE SPARK PLUG.

---CAUTION---

A. **Throttle Plate and Throttle Shaft Assembly**
B. **Idle Speed Regulating Screw**
C. **Idle Fuel Regulating Screw**
D. **Choke Plate and Choke Shaft Assembly**
E. **Atmospheric Vent Hole**
F. **Float Bowl Housing Drain Valve**
G. **Float Bowl Housing Retainer Screw**
H. **High-speed Adjusting Needle**
J. **Float Bowl Housing**
K. **Idle Chamber**

FIGURE 215   Side Draft, Float Bowl Type Carburetor

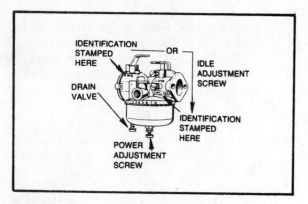

FIGURE 216   Type LMG Carburetor

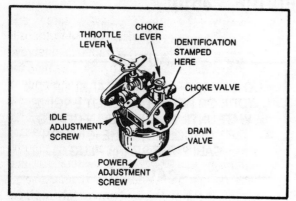

FIGURE 217   Type LMB Carburetor

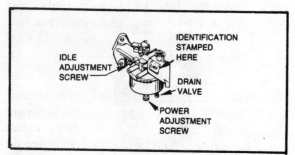

FIGURE 218   Type LMV Carburetor

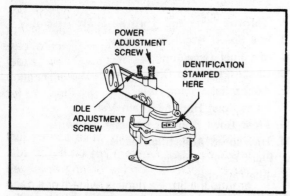

FIGURE 219   Type HEW Carburetor

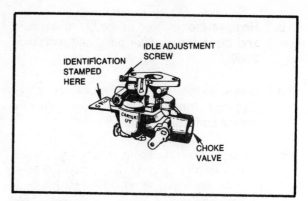

FIGURE 220A   Type UT Carburetor

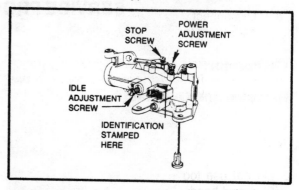

FIGURE 220B   Suction Lift Type Carburetor

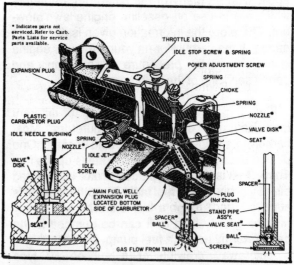

FIGURE 221   Cross Section of Typical Junction Lift Type Carburetor

CARBURETOR REMOVAL

Step 1—

a.   Shut off fuel supply valve. If this cannot be done, or there is no shut off valve, drain the fuel tank.

b. Remove the air filter assembly or air filter, and clean the air filter per Fault Indication **2140.**

c. Clean away all dirt, grease or grass clippings from the carburetor and the engine surfaces.

d. Carefully note, and write down, the location and position of all external linkages, cables, clamps, etc., before detaching them from the carburetor.

e. Disconnect the fuel supply line to the carburetor (if there is a supply line).

f. Make certain to use the exact size wrench or screwdriver to loosen the bolts or screws which hold the carburetor assembly to the engine.

NOTE:

Do not use too small a screwdriver for bolts with slotted heads. Use the proper fitting wrench. Damaged bolt heads or screw slots will make reinstallation difficult or even impossible.

g. Lift the carburetor from the engine. Take note of any gasket which is between the carburetor end and the engine mounting pad. Remove the gasket, and make certain the engine mounting pad is clean and free of gum or dirt.

h. Proceed to Step 2.

Step 2—Visually examine the carburetor assembly. Look for defects, wear or broken parts as follows:

a. Examine the throttle shaft. The shaft may be rubbing or binding and in so doing has enlarged the shaft hole or has rubbed or worn a hole through the carburetor throat.

b. Examine the carburetor casing for cracks or pieces broken away.

Are both (or any one) of these faults evident?

YES—The carburetor cannot be serviced and must be discarded for a new exact replacement. Refer to Figs. 215 to 220 for locations of carburetor manufacturers and identification markings or model number. Proceed to Step 7 for installation of new carburetor.

NO—Proceed to Step 3.

Step 3—Disassemble the carburetor as follows. Place the carburetor on a clean surface or clean cloth. Note carefully the use and location, of all gaskets removed. Refer to the exploded line drawings of Figs. 222 to 230.

a. Remove the idle adjustment screw and screw spring (or needle), provided there is one on the carburetor. Remove by unscrewing by hand, counterclockwise. Examine the needle screw mounting hole in the body of the carburetor. Look for evidence of wear at the point where the end, or cone part of the needle, seats in the hole. If the hole has worn down into the same shape as the end of the needle or has been enlarged so that the needle fits loosely, the needle seat is worn. If the seat is not separately replaceable, the carburetor body is no longer serviceable. Discard the carburetor and replace with a new exact replacement. See figures cited under Step 2 for identification markings and proceed to Step 7 for installation.

b. If the needle seat passes inspection remove the main (high speed or power) adjustment needle and needle spring (also called adjustment screw) provided there is one on the carburetor. Remove by unscrewing by hand, counterclockwise. Examine the mounting hole as per Step 3a. If this mounting is worn and if the seat is not separately replaceable then discard carburetor and replace with a new exact replacement. See Step 2 for identification markings and proceed to Step 7 for installation.

c. Examine the two needle valve screws just removed. Look for worn tips on the needle valves. Look for a ridge or ring engraved around the tip. If a ridge is evident or a deep ring, then the needle valve is defective. Set aside and proceed.

d. Examine the throttle valve butterfly inside the throat and if this valve can be removed by removal of its fasteners, remove the valve. Do the same for the choke valve butterfly and remove the valve if there is one in the carburetor. Remove any internal metering jets or screws. Locate and remove the screws which hold the body of the carburetor together. Remove the screws. Place the carburetor parts on the clean surface and remove each piece and part as shown in the typical exploded line drawing for the carburetor in figures cited under Step 3.

e. Examine each individual part. Look for worn parts (as the ridges or rings around the needle valves), clogged up ports or holes, and internal surfaces and passages which have gum or dirt deposits.

f. Place all gaskets aside. All gaskets must be replaced with new gaskets even if in apparently good condition.

g. For carburetors equipped with float valves as shown in the typical float valve assembly of Fig. 231, examine the float for leaks, holes or cracks. If the float is made of cork and sealed externally with a sealer, look for loss of the sealant on the surface. Floats which are faulty will cause irregularities in the flow of fuel into the carburetor bowl or reservoir. Examine the float bracket, pin, and hinge for signs of wear.

NOTE:

If any of the float assembly parts are found to be faulty, discard the complete float assembly. Replace with a new exact replacement. Do not attempt to replace individual components as an improperly positioned part will not sit properly and will cause problems which will be hard to isolate.

h. Proceed to Step 4 for cleaning the disassembled parts.

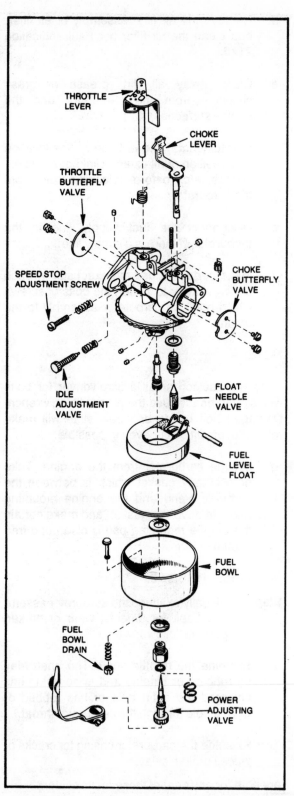

FIGURE 222   Exploded View, Typical Float and Needle Valve Type LMG, LMB, LMV Carburetor

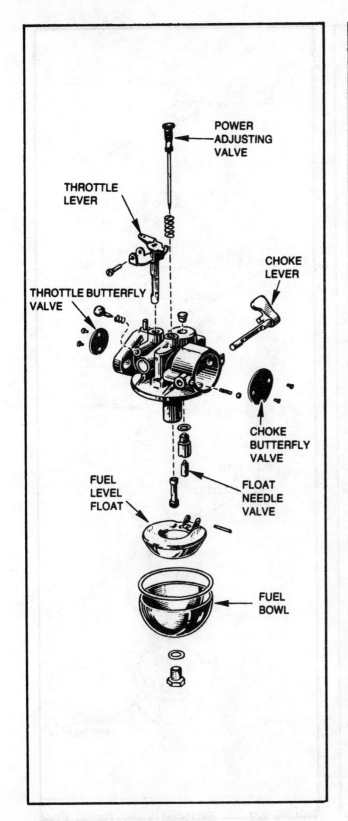

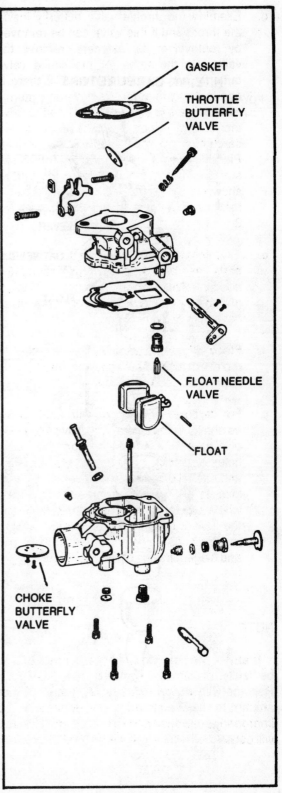

FIGURE 223 Exploded View, Typical Float and Needle Valve Carter Carburetor

FIGURE 224 Exploded View of UT Carburetor

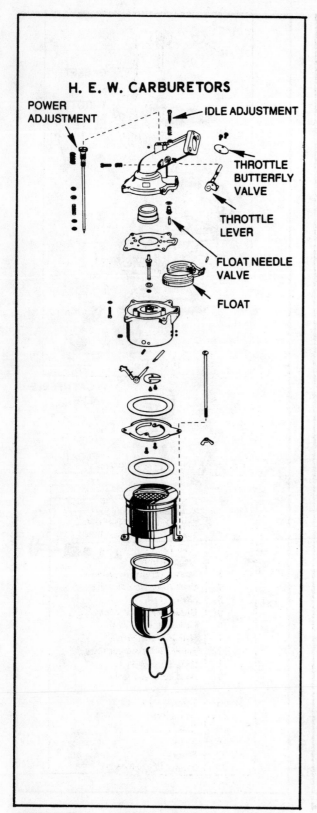

H. E. W. CARBURETORS

POWER ADJUSTMENT

IDLE ADJUSTMENT

THROTTLE BUTTERFLY VALVE

THROTTLE LEVER

FLOAT NEEDLE VALVE

FLOAT

FIGURE 225   HEW Carburetor, Exploded View

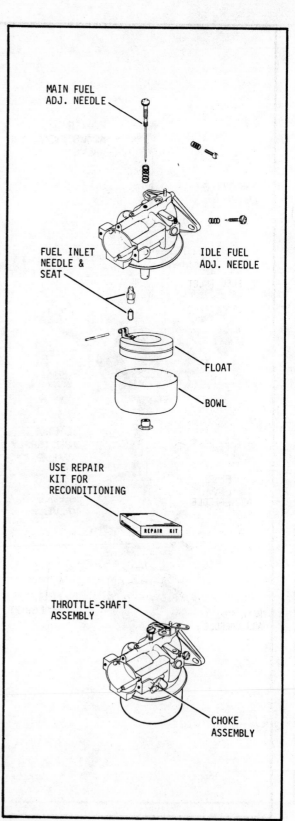

MAIN FUEL ADJ. NEEDLE

FUEL INLET NEEDLE & SEAT

IDLE FUEL ADJ. NEEDLE

FLOAT

BOWL

USE REPAIR KIT FOR RECONDITIONING

REPAIR KIT

THROTTLE-SHAFT ASSEMBLY

CHOKE ASSEMBLY

FIGURE 226   Exploded View Side Draft Carburetor

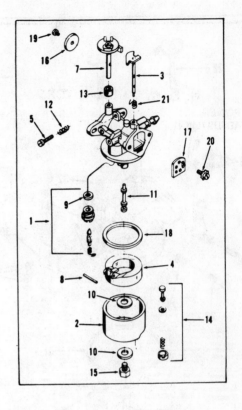

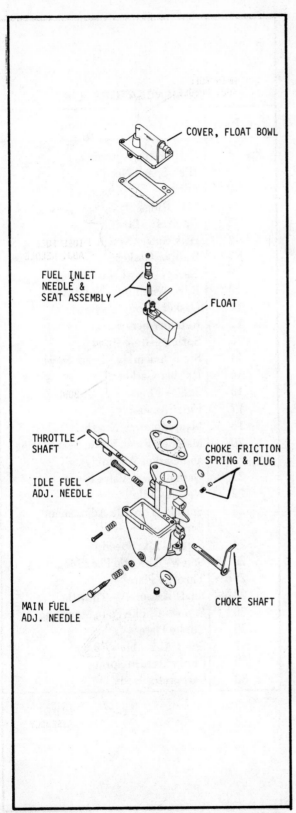

NOMENCLATURE LIST (FIXED JET)

Fig - Item		Description
228 -	1	Matched, Float Valve Seat, Spring and Gasket Assembly
	2	Bowl Assembly - Float Bowl
	3	Shaft Assembly - Choke
	4	Float Assembly
	5	Screw - Throttle Adjustment
	7	Shaft Assembly - Throttle
	8	Shaft - Float
	9	Gasket - Float Valve Seat
	10	Gasket - Nut To Bowl
	11	Main Metering Nozzle
	12	Spring Throttle Adjustment
	13	Seal - Throttle Shaft
	14	Bowl Drain Assembly
	15	Retainer Screw
	16	Throttle Plate
	17	Choke Plate
	18	Gasket - Bowl To Body
	19	Screw
	20	Screw
	21	Spring - Choke Return

FIGURE 227   Exploded View Up Draft Carburetor

FIGURE 228   Carburetor, Exploded View

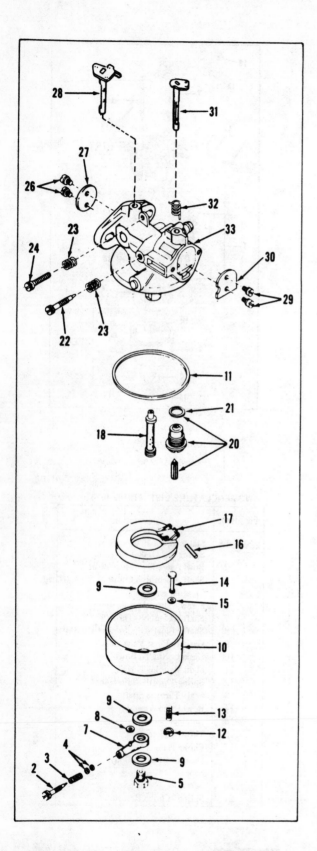

NOMENCLATURE LIST

Fig - Item	Description
229 · 1	Carburetor Assembly
2	High-Speed Needle
3	Spring
4	"O" Rings
5	Retainer - Bowl
7	High-Speed Needle Housing
8	Rubber Gasket
9	Gasket - Bowl Nut To Bowl
10	Bowl Assembly - Float Bowl
11	Gasket - Body To Bowl
12	Retainer Screw
13	Spring - Drain Bowl
14	Stem Assembly - Drain Bowl
15	Rubber Gasket
16	Shaft - Float
17	Float Assembly
18	Main Metering Nozzle
20	Matched Float Valve, Seat, Spring and Gasket Assembly
21	Gasket - Seal Valve Float
22	Needle - Idle
23	Spring - Throttle Adjustment Screw
24	Screw - Idle Speed
26	Screw - Throttle Plate Mounting
27	Throttle Plate
28	Shaft Assembly - Throttle
29	Screw - Choke Plate Mounting
30	Choke Plate
31	Shaft Assembly - Choke
32	Choke Return Spring
33	Carburetor Body

FIGURE 229   Carburetor, Exploded View

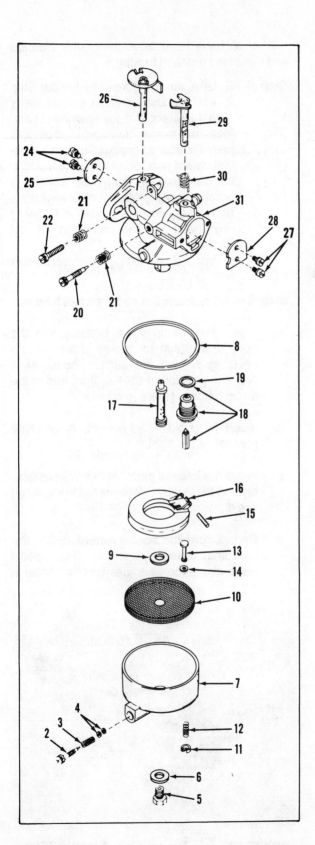

NOMENCLATURE LIST

Fig - Item	Description
230- 1	Carburetor Assembly
2	High-Speed Needle
3	Spring
4	"O" Rings
5	Retainer Screw
6	Gasket - Bowl Nut To Bowl
7	Bowl Assembly - Float Bowl
8	Gasket - Body To Bowl
10	Screen
11	Retainer Screw
12	Spring - Drain Bowl
13	Stem Assembly - Drain Bowl
14	Rubber Gasket
15	Shaft - Float
16	Float Assembly
17	Main Metering Nozzle
18	Matched Float Valve Seat, Spring and Gasket Assembly
19	Gasket - Seal Valve Float
20	Needle, Idle
21	Spring
22	Screw, Throttle Adjustment
24	Screw - Throttle Plate Mounting
25	Throttle Plate
26	Shaft Assembly Throttle
27	Screw - Choke Plate Mounting
28	Choke Plate
29	Shaft Assembly - Choke
30	Spring Choke Return
31	Carburetor Body

FIGURE 230   Carburetor, Exploded View

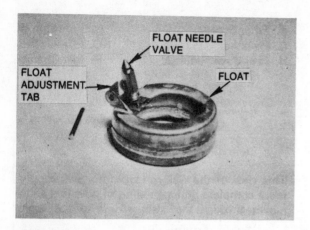

FIGURE 231   Typical Carburetor Float Valve
Assembly

— CAUTION —

USE MAXIMUM VENTILATION
WHEN USING CARBURETOR CLEANER.
FOLLOW MANUFACTURER'S DIRECTIONS.
FUMES ARE FLAMMABLE AND HARMFUL.

—CAUTION—

Step 4—Soak all the parts in a carburetor cleaner or solvent. Use a clean container in a well ventilated room. Inspect the parts to insure they are clean and free of all dirt and gum. This is especially important for the barrel or throat area of the carburetor.

Remove each piece, let dry, and reinspect to be sure it is clean. If not, resoak in cleaner. Blow out all the small carburetor passages. A simple bicycle or ball pump is helpful for this task.

NOTE:

Do not use a brush or lint cloth or towel to clean these passages. The passage can be enlarged by use of a brush. Lint or hair may lodge in passage. This will distort any accurate setting.

After all parts have been thoroughly cleaned and inspected proceed to Step 5.

Step 5—Refer again to figures cited under Step 2, which show locations of carburetor manufacturers and identification markings or model numbers. Use this identification to procure the exact replacement part for all the parts found to be defective or faulty. Be certain to procure a complete set of new gaskets (they usually are available in kit form) regardless of the condition of old gaskets.

Proceed to Step 6 for carburetor reassembly.

Step 6—Reassemble the carburetor as follows.

a. Insert the throttle valve butterfly into the body and attach to throttle shaft with the securing screws. Be certain the valve is properly positioned on the shaft and in the throat before tightening screws.

b. Insert the choke valve butterfly in the same manner as the throttle.

c. Install the internal parts, gaskets, jet assembly and the like by reference to the exploded line drawings.

d. Reassemble the float assembly/bowl. Before final assembly, set the float needle valve clearance or adjustment as shown in

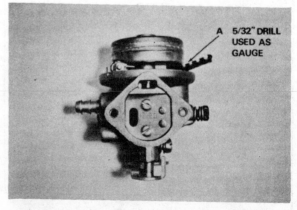

FIGURE 232   Typical Float Assembly Clearance Setting

the typical float assembly of Fig. 232. Refer to Table V for a listing of typical carburetor float valve clearance values, or use your individual engine manufacturer's specifications. Adjustment to the float valve clearance is performed by inverting the carburetor and assembled float valve as indicated in Fig. 232. Using a drill bit of the correct diameter, gives an excellent gage for the clearance measure of the float to the outer rim of the body casing. Bending the float tab or tang permits an exact adjustment.

TABLE V

LISTING OF TYPICAL CARBURETOR MANUFACTURERS' MODEL NUMBERS AND FLOAT VALVE SETTINGS

Carburetor Manufacturer and Model Number	Float Setting in inches	Initial Idle and Power Screw Needle Adjustments, Number of Turns Open	
		Idle	Power
Carter NS Series		Turns	Turns
N705S, N707S	13/64	1½	1½
N2003S, N2029S	13/64	1	3/4
N2020S, N2087S	11/64	1½	1½
N2147S, N2171S	11/64	1½	1½
N2236S, N2246S	11/64	1	1½
N2399S, N2019S	11/64	1½	2
N2457S, N2449S	13/64	1½	1½
N2456S	13/64	1½	1½
N2459S	11/64	1	1½
N2466S	11/64	1½	2
Carter UT Series			
2217S	11/64	1	1
2230S	17/64	1	1
2336S, 2337S, 2398S	1/4	1	1
2712S, 2713S	19/64	1	1
2714S	1/4	1	1
Carter HEW Series	3/16	1½	1½
Marvel-Schebler			
AH22A, AH29	3/32	1	1¼
Walbro			
LMG, LMB, LMV	5/32	1¼	1¼

NOTE:

Bend the tab in very small increments and check setting after each increment. Do not bend the tab if there is a mechanical screw adjustment provided for this purpose.

Install the bowl ring gasket, and then the bowl cover. Make certain that the idle jet is not blocked by the ring gasket.

Step 7—Install the gasket (if one is used) between the carburetor and the engine, and install carburetor to engine with mounting bolts or screws. Tighten securely.

Step 8—Attach all the external linkages previously removed making sure to follow the markings, or record made earlier, of the positions of linkages, cables or clamps.

Step 9—Install the needle assembly valves, and springs. Be certain to replace the idle valve into the idle valve hole and likewise the power valve into the power hole. Use only hand pressure to screw these valves in.

NOTE:

Do not force these needle valves in and do not screw down tight. Use only the fingers since heavy force will damage valves.

Step 10—Using only the fingers, carefully screw the power adjusting valve in until it is fully seated and finger tight. After it is seated, back the screw out 1 to 1½ turns. Do the same for the idle adjusting screw, seat it finger tight, then back it out 1 turn. Subsequent adjustment for idle and maximum power RPM (speed) will be made with the use of a Tachometer.

Step 11—Connect the ignition cable to the spark plug. In Step 10 a gross adjustment was made for the carburetor idle and power needle screws. It is necessary to do a fine tuning for this adjustment with the engine running. Proceed to Step 12.

Step 12—This step provides a simple procedure for adjusting engine speed by a direct reading of engine RPM (revolutions per minute).

A simple engine Tachometer is necessary. A Tachometer can be purchased at any automobile supply store, larger department store, and at most electronic supply shops. An inexpensive Tachometer is fully sufficient for this task.

It will be necessary to locate the hot point or hot lead wire on the engine. On most engines the hot lead wire is the smaller diameter of the two wires which come out from under the magneto flywheel. The heavy wire is the high tension ignition cable. This goes to the spark plug. The other wire is the hot lead wire and it usually goes to the starter switch, or to the control lever, or to a point near the carburetor where movement of the control cable to the STOP position shorts the hot lead (voltage) to ground and stops the engine. Refer to Figs. 233, 234 and 235.

Connect up the Tachometer as shown.

a. The "hot point" is that point where the primary wire of the magneto coil is connected along with the condenser wire, and the breaker points. It is from this common point that the hot lead wire goes to the starter switch or control cable short-out location. The black or ground lead to the Tachometer can be connected to any convenient good ground point, but using the base of the spark plug is the preferred point.

b. Refer to the engine manufacturer's specification for the idle speed RPM and top speed RPM, or Service Procedure **4050.** If this information is not available, then a fair value for idle RPM speed is about 1300 or 1400 RPM. Top speed is about 3500 RPM. Proceed to adjust the speed as follows.

c. Start up engine. Let engine warm up at IDLE control lever position about 5 minutes. Meanwhile, read the RPM on the Tachometer.

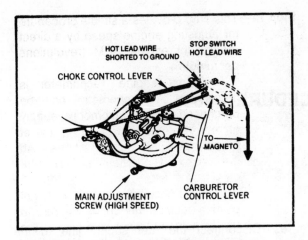

FIGURE 233 Typical Location of Hot Lead Stop Switch Wire-Engine Control in "Stop" Position

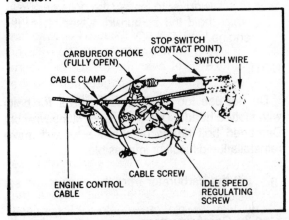

FIGURE 234 Typical Positions of Control Cable for Fast or High Speed Position

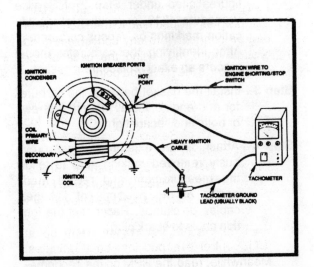

FIGURE 235 Tachometer Hook-up Connections

NOTE:

In the event the engine does not start up, check the settings of the carburetor idle and power needle valves. It may be necessary to make small adjustments to these settings if the engine does not start.

d. Using your fingers, adjust the *idle* until the desired idle RPM is reached as measured on the Tachometer.

e. Reposition the control lever to maximum or high speed. Let engine run at this speed for a few minutes.

f. Using your fingers adjust the power screw until the desired RPM is reached on the Tachometer.

g. With the Tachometer connected, move the control back and forth from idle to high speed. The engine should respond smoothly and should stay at the speed to which the control lever is positioned. The idle RPM and top speed RPM should be those for which you adjusted the carburetor. If not, some fine tuning by small adjustments may be necessary.

NOTE:

In the event that the desired RPM values cannot be obtained, shut down the engine and re-examine the needle valves (adjusting screws). The screws may have been damaged in the assembly or adjustment operation. If so, replace with new screws. If the desired RPM valves cannot be reached and are far out of line (say 100 to 200 RPM or more) with the recommended settings, then the valve seals in the carburetor body may be worn out. In this case a new carburetor is required.

SERVICE PROCEDURE   4011

Carburetor Assembly Removal for Replacement of Leaky Cylinder Gasket

Before doing any work on the carburetor the following caution should be strictly observed.

```
┌──────────── CAUTION ────────────┐
│                                  │
│        DO NOT PERFORM ANY        │
│   CARBURETOR WORK ON A HOT       │
│   ENGINE. WAIT UNTIL ENGINE      │
│       HAS FULLY COOLED.          │
│                                  │
└──────────── CAUTION ────────────┘
```

Step 1—Remove the carburetor assembly from the engine as follows:

a. Shut off fuel supply valve. If this cannot be done, or there is no shut off valve, drain the fuel tank.

b. Remove the air filter assembly or air filter, and clean the air filter per Fault Symptom **2140**.

c. Clean away all dirt, grease or grass clippings from the carburetor and the engine surfaces.

d. Carefully note, and write down, the location and position of all external linkage, cables, clamps, etc. before detaching them from the carburetor.

e. Disconnect the fuel supply line to the carburetor (if there is a supply line).

f. Make certain to use the exact size wrench or screwdriver to loosen the bolts or screws which hold the carburetor assembly to the engine.

NOTE:

Do not use too small a screwdriver for bolts with slotted heads. Use the proper fitting wrench. Damaged bolt heads or screw slots will make reinstallation difficult or impossible.

g. Lift the carburetor from the engine and set aside.

Step 2—Carefully clean away all of the old gasket from the engine mounting pad and from the carburetor flange. See figures cited under Step 2 of Service Procedure 4010 for location of identification markings on various carburetors. After identifying the carburetor model procure an exact replacement gasket.

Step 3—Install the gasket between the carburetor and engine, install mounting screws or bolts and secure tightly.

Step 4—Attach all the external linkages previously removed making sure to follow the measurement and record made earlier of the positions of linkages, cables or clamps. Reconnect the ignition cable to spark plug.

SERVICE PROCEDURE   4012

Carburetor Removal for Replacement of Leaky Reed Valve

The following caution should be strictly observed.

```
┌─────────── CAUTION ───────────┐
│                                │
│      DO NOT PERFORM ANY        │
│   CARBURETOR WORK ON A HOT     │
│   ENGINE. WAIT UNTIL ENGINE    │
│      HAS FULLY COOLED.         │
│                                │
└─────────── CAUTION ───────────┘
```

REMOVAL OF CARBURETOR ASSEMBLY

a. Shut off fuel supply valve. If this cannot be done, or there is no shut off valve, drain the fuel tank.

b. Remove the air filter assembly or air filter, and clean the air filter per Fault Indication **2140.**

c. Clean away all dirt, grease or grass clippings from the carburetor and the engine surfaces.

d. Carefully note, and write down, the location and position of all external linkage, cables, clamps, etc., before detaching them from the carburetor.

e. Disconnect the fuel supply line to the carburetor (if there is a supply line).

f. Make certain to use the exact size wrench or screwdriver to loosen the bolts or screws holding the carburetor assembly to the engine.

NOTE:

Do not use too small a screwdriver for bolts with slotted heads. Use the proper fitting wrench. Damaged bolt heads or screw slots will make reinstallation difficult or even impossible.

g. Lift away the carburetor assembly. The carburetor and reed adaptor should be removed as a unit if possible, see Fig. 236.

h. Remove the reed adaptor assembly from the engine after removal of the mounting screws. Lift the reed adaptor or plate out of the engine. Note how the valve was installed in the engine. See Figs. 237 and 238 for typical reed valves and adaptors.

FIGURE 236   Typical 2-Cycle Engine Carburetor Nomenclature

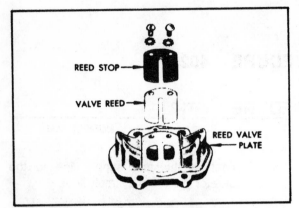

FIGURE 237   Typical Reed Valve and Adapter Plate

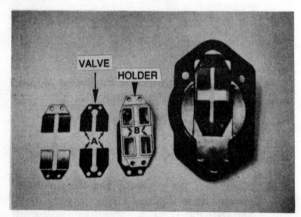

FIGURE 238A   Typical Reed Valve and Adapter

i. Examine the adaptor for proper reed valve seating on the assembly. The reed valve seats properly when the reeds seat with no clearance. Replace obvious defective reed valves which are bent or broken. The reeds are pre-set to create a pressure side, therefore attempts at straightening are useless.

j. Procure new reed valves which are exact replacements. Be sure to install in the same position as the old valves. The pressure side must always face downward. If the die cast part of the valve is warped, replace with a new one.

k. Reassemble the valve in reverse order of disassembly of Step h, being certain the valve is positioned properly. Refer to exploded line drawing figures given in **4001** for assembly order.

l. Install a new reed adaptor mounting gasket and secure plate assembly to engine. Install a new carburetor mounting gasket and install carburetor. Install mounting screws or bolts and tighten.

m. Attach all the external linkages previously removed making sure to follow the measurement and record made earlier of the positions of linkage, cables or clamps. Reconnect the ignition cable to spark plug.

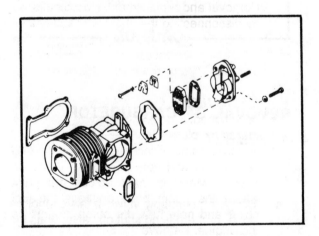

FIGURE 238B   Typical Reed Valve Assembly

SERVICE PROCEDURE 4020

Fuel Pump, Removal and Overhaul

Most fuel pumps used on lawn mowers are one of two types, mechanical or vacuum. The mechanical pump is operated by means of an eccentric or a cam on the engine's crankshaft. Typical mechanical pumps are shown in the exploded parts view of Fig. 239 and the cross-sectional view of Fig. 240. The vacuum pump is operated by the pressure within the crankcase which rises and falls and is sufficient to drive a diaphragm inside the pump. Fig. 241 shows a cross-section of a typical vacuum pump.

- *VACUUM TYPE FUEL PUMPS* cannot be repaired and if defective should be replaced with a new exact replacement pump. Removal and replacement is accomplished by disconnecting the impulse line, fuel inlet and outlet lines and unbolting from the carburetor. Replacement of a new pump is accomplished by reversal of this order.

- *MECHANICAL FUEL PUMPS* can be reconditioned. Remove the ignition cable from the spark plug. Clean all the area around the fuel pump. Disconnect fuel inlet and outlet lines and note position of pump on engine. Make a mark across the paint where the pump body engages the fuel cover and note how the pump mounts on the engine. Remove the pump by unbolting from engine. Note the position of the mounting gasket to be sure to replace in same position. Refer to the exploded parts diagram of the pump in Fig. 239.

a. Be certain to mark the pump body and cover as noted.

b. Place pump on a clean area or cloth to disassemble.

c. Remove cover screws and set aside in can or pan. Remove the pump cover carefully.

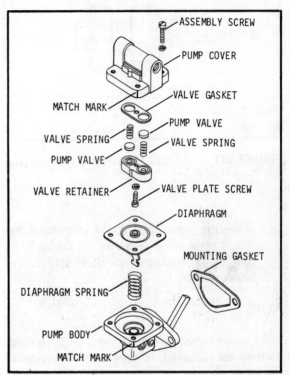

FIGURE 239   Mechanical Type Fuel Pump, Exploded View

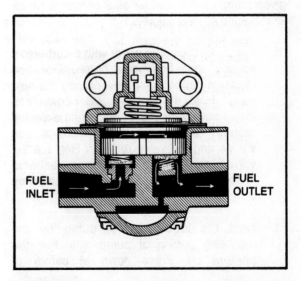

FIGURE 240   Cross Section of Mechanical Fuel Pump

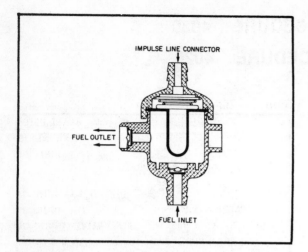

FIGURE 241 Cross Section of Typical Vacuum Type Fuel Pump

d. Turn the cover over so as to remove the valve plate screw and washer. Carefully lift out the valve retainer, the valve springs and valve gasket.

NOTE:

Be sure to observe the positions of these parts as they are removed, so they can be replaced in the same manner.

Discard the valve springs, valves and valve retainer gasket. Procure a new set of these parts. (Usually these parts can be purchased for the given pump model number as a complete repair kit including all new gaskets.)

e. Clean out the pump cover with a carburetor solvent and fine wire brush. Dry and reassemble the pump cover using the new parts. Hold the cover upside down and install new valve gasket. Be sure it is seated properly all around. Install springs and valves and the valve retainer. Replace the valve plate screw; be certain the washer is in place under the screw head. Place pump cover aside.

f. Hold the lower part of pump by the mounting portion of pump, with the diaphragm up. Press down at center of diaphragm and while pressing, rotate the diaphragm ¼ of a turn. This will unhook the diaphragm. Remove diaphragm and spring.

g. Clean out the body of the pump with carburetor cleaner and a fine wire brush. Let dry.

CAUTION

USE MAXIMUM VENTILATION WHEN USING CARBURETOR CLEANER. FOLLOW MANUFACTURERS' DIRECTIONS.
FUMES ARE FLAMMABLE AND HARMFUL.

CAUTION
---

h. Hold the pump body and install diaphragm spring in an upright position. Examine the diaphragm carefully. Look for frayed edges, small holes or tears. Discard, if any of these are found, and replace with a new diaphragm. Place diaphragm tang into the spring and press down gently on diaphragm to compress the spring. With spring compressed, rotate the diaphragm ¼ of a turn so as to reconnect the diaphragm. Test that the diaphragm is seated by *gently* pressing down on the center and releasing quickly. The diaphragm should stay seated.

i. Assemble the pump cover to the lower body. Make certain the alignment mark made earlier is lined up. Replace the screws *but do not tighten them.* Using your hand, push the pump lever down as far as it will travel, and then tighten up the screws. This operation will keep the diaphragm from stretching.

j. Reinstall the pump on the engine using a new mounting gasket. Connect the inlet and outlet fuel lines.

k. Reconnect the ignition cable to the spark plug.

SERVICE PROCEDURE   4030

Ignition System

Overhaul of the ignition system should be conducted on the basis of test and replacement of those parts which require the most frequent adjusting, or which are known to wear out under normal use. The spark plug is the first part which should be serviced and is covered in Service Procedure **4004.** The next part is the set of ignition breaker points. With the engine shut down remove ignition cable from spark plug, and then remove the plug as described in Service Procedure **4004.**

• *IGNITION BREAKER POINTS, EXTERNAL POINT SYSTEM.* The ignition system on some lawn mower engines such as Kohler, shown in Fig. 242, feature breaker points which are on the outside of the engine and can be serviced without removal of the engine flywheel magneto. If examination shows the breaker points, see Fig. 243, are burned, pitted, worn out or dirty, proceed as follows for point set replacement and proper point gapping.

Step 1—Remove the breaker point cover and cover gasket. Handle the gasket carefully so that it can be reused. Place screws and gasket in a container for safekeeping.

Step 2—Refer to Fig. 242 and slowly turn or crank the engine until the breaker points are at the maximum opening or gap. This should occur when the push rod (or cam) is at its maximum travel or distance out of the engine. DO NOT DISTURB OR CRANK THE ENGINE UNTIL NEW POINTS ARE INSTALLED AND GAPPED.

Step 3—Unscrew lead wire from points. Unscrew retaining screws on points set and remove the points from engine.

Step 4—Install new point set using the same exact replacement parts. Screw down the hold down screw fasteners but leave the adjusting lock screw (as shown in Fig. 243) just snug, not tight.

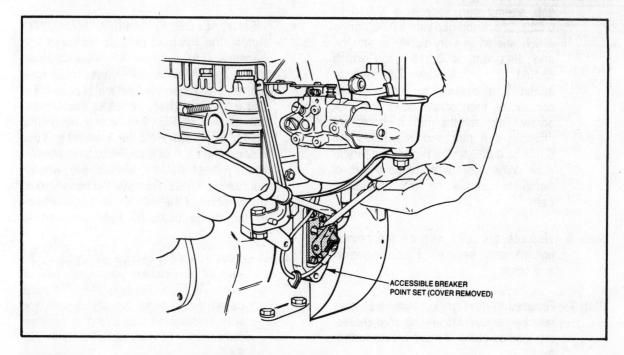

ACCESSIBLE BREAKER
POINT SET (COVER REMOVED)

FIGURE 242   Typical Accessible Breaker Point Set

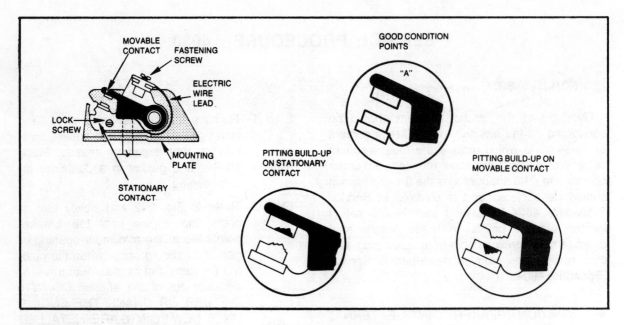

FIGURE 243   Good Condition Breaker Points and Poor Condition Pitted Points

Step 5—Refer to Fig. 242. See Table V (a) for point gap setting. The Kohler engine is being used in this example. Using a 0.020 inch thick feeler gage, insert a screwdriver into the adjusting slot and set the point gap to 0.020 inches. This gap setting can vary from 0.018 to 0.022. Check the gap with a 0.022 gage which should fit very tightly or snugly, and also with a 0.018 gage which should fit very loosely. Continue to adjust if necessary to obtain these conditions, then screw down the lock screw. Recheck the gap. It should not change; if it has and is outside the 0.018 to 0.022 limits, repeat the procedure. When complete, place a dab of lubriplate grease on the crankshaft cam.

Step 6—Replace the lead wire on the points, tighten screw securely. Replace gasket and cover.

Step 7—Proceed to next ignition system step as required or operate engine after replacing ignition cable, if there are no farther faults.

• *IGNITION BREAKER POINTS, INTERNAL POINT SYSTEM.* Internal breaker points are located under the flywheel magneto, and therefore require the removal of the flywheel to reach and service them. Proceed to Step 4, and then to Steps 5 and 6.

• *REMOVAL OF FLYWHEEL MAGNETO.* Before the flywheel can be removed it is necessary to remove all encumbrances such as shrouds, air vane deflectors, starter, etc. When the flywheel is accessible use a flywheel strap, or holder such as the one shown in Fig. 244, and position the holder as shown. Hold the flywheel and use a socket wrench or a box wrench to remove the flywheel nut by turning the wrench counterclockwise. Remove the nut and note the position of the washer so as to replace in the same position. Then proceed as follows:

Step 1—Most flywheels can be removed by the use of a flywheel knock-off nut as shown in Figs. 245 and 246. If in some cases the flywheel cannot be removed with a knock-off nut, use of a flywheel puller is necessary as shown in Fig. 247.

FIGURE 244   Holding Flywheel by Use of Strap

FIGURE 247   Removal of Flywheel with Flywheel Puller

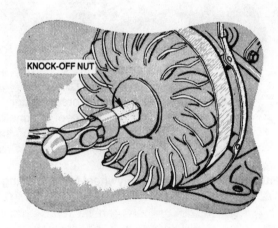

FIGURE 245   Typical Flywheel Removal with Knock-off Nut

FIGURE 246   Knock-off Nut Used in Flywheel Removal

Step 2—DO NOT USE THE ORIGINAL FLYWHEEL NUT AS A KNOCK-OFF NUT. This will damage the crankshaft threads.

Step 3—Place the knock-off nut on the crankshaft as shown in Figs. 245 and 246, and strike the nut sharply with a heavy ballpeen (or other heavy) hammer.

Step 4—The flywheel will pop off (or will be pulled off if the flywheel puller is used).

• *FLYWHEEL INSPECTION AND TEST.* Inspect the flywheel closely. Look inside the flywheel for evidence of interference between the flywheel and the stator assembly. Scraping and damage to both the stator and flywheel will result from such interference. Also examine the condition of the flywheel key, and the keyway slot. If the key is worn or has burrs, discard and replace with a new exact replacement key. Check the taper on the crankshaft. If the shaft has worn, there will be bright marks on the shaft taper. Both the taper hole in the flywheel and the crankshaft taper should have completely dull surfaces. See Fig. 248. If the flywheel is suspected of losing its magnetic strength the following simple test can be run on the magnet.

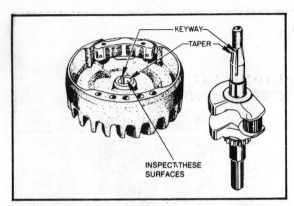

FIGURE 248 Flywheel and Crankshaft End Inspection (Crankshaft Shown Removed from Engine)

Step 1—Place a ½ inch socket on the magnet inside the flywheel as shown in Fig. 249.

Step 2—Now shake the flywheel back and forth and sideways.

Step 3—If the socket cannot be held inside the flywheel by the force of the flywheel magnet then the magnet does not have sufficient strength. Replace the flywheel with a new exact replacement part.

NOTE:

Hammering on, or dropping, the flywheel can lead to loss of magnetic strength of the magnet.

Step 4—If a ½ inch socket is not available use a non-magnetic screwdriver to test for magnetic strength. Hold the blade of the screwdriver about one inch from the magnets; the magnets should pull the blade into them. If not, replace flywheel unit with a new one.

● *IGNITION BREAKER POINTS RE-PLACEMENT, INTERNAL BREAKER POINT SYSTEM.* With the flywheel removed, inspect the stator or ignition coil assembly. Take note of the position and location of the high tension ignition cable and the ignition wire, etc. Refer to Fig. 250 in which the breaker point dust cover is removed. Remove/replace the points as follows:

Step 1—Slowly turn or crank the engine until the breaker points are at the maximum opening or gap. This should be when the cam on the crankshaft has its highest or thickest point in contact with the movable arm of the points. DO NOT DISTURB OR CRANK THE ENGINE UNTIL NEW POINTS ARE INSTALLED AND GAPPED.

Step 2—Unscrew and remove the screw holding the wire going from the points to the point/condenser terminal. Unscrew and remove the hold down screw.

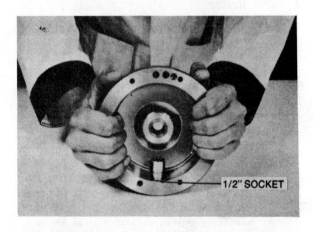

FIGURE 249 Test of Flywheel Magnet Using 1/2" Socket

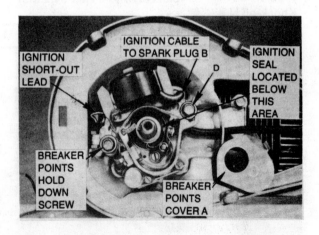

FIGURE 250 Engine Flywheel Magneto Removed Showing Ignition System Components and Location of Ignition Seal

TABLE V(a)

TABULATION OF BREAKER POINT GAP
SETTINGS AND TOLERANCES FOR SELECTED LAWN MOWER ENGINES

Engine Manufacturer and Model Number	Point Gap Setting in Inches	Setting Tolerance or Range in Inches
Jacobsen Model 321	0.020	—
Kohler K91, K141, K161, K181, K241, K301, K321	0.020	0.018–0.022
Clinton Series 1600, 1600–1000, 1800–1000, 2500, A2500, B2500–1000, 2790–1000 E10–1000, 414–1300–000 414–1301–000, 416–1300–000, 418–1300–000, 418–1301–000, 420–1300–000, 420–1301–000, 422–1300–000, 422–1301–000	0.029	0.028–0.030
E10–1000	0.015	0.013–0.017
E75–1000, E95–1000	0.015	0.014–0.016
All other series	0.020	0.018–0.021
Lawn Boy	0.020	—
Briggs and Stratton All Models	0.020	—
Tecumseh AH31, AV31, AH36, AV36, AH80, AV80	0.019	0.018–0.020
AV47, AH81, AV81, AH82	0.018	0.015–0.021
AH47, AH58, AV58, AH61, AV61 , AH440, AH480, AH490, AH750	0.017	0.015–0.019
AH51, AV51, AH520, AV520	0.020	0.017–0.023
AV600, AV750	0.018	0.016–0.020
All Others	0.020	—

Step 3—Carefully lift the point set out of the engine. Replace with a new exact replacement set. Fasten in position and attach the condenser lead to screw; tighten screw.

Step 4—Refer to Figs. 251 and 252. Using a 0.020 inch thick feeler gage, move the fixed plate of the point set and set the point gap to 0.020 inches (or as called out by engine manufacturer - see Table V). This gap setting can vary one or two thousands up and down (0.019 to 0.021). Lock the points into position by tightening the fixed point arm lock screw. Check the gap with a 0.021 gage, which should fit very tightly or snugly and then with a 0.019 gage, which should fit loosely. Continue to adjust if necessary to obtain the manufacturer's recommended point gap setting. When complete, place a dab of lubriplate grease on the crankshaft cam.

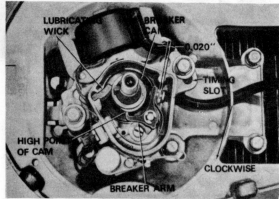

FIGURE 251 Typical Set of Ignition Breaker Points Installed in Engine

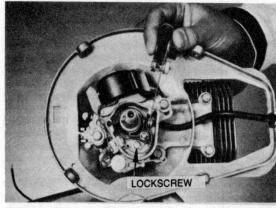

FIGURE 252 Setting the Breaker Point Gap

NOTE:

This step has used a breaker point gap setting of 0.020. Refer to your individual engine manufacturer's specification for your particular point gap setting. If unavailable use Table V as a reference.

• *IGNITION TIMING.* Upon completion of points replacement the timing of the engine should be checked. The timing of the engine determines the point at which the breaker points begin to open and the spark plug is fired. Timing should be set in accordance with the engine manufacturer's specification. Timing varies for different engines from before the piston reaches the top of its stroke to after the top of the stroke. This is usually referred to as before top center, BTC, to after top center, ATC, or even at top center, TC. Improper timing can result in the engine not developing its full power capability. There are many variations in timing procedures for different engines. Two typical procedures are given which serve as a guide for most mower engines.

Step 1—*Timing slot on breaker magneto stator plate.* Fig. 253 shows a standard breaker magneto assembly removed from the engine. Figs. 254 and 255 show a breakerless magneto and solid state breakerless magneto—both of which are timed in the same manner as the standard breaker magneto. The two elongated mounting holes on the stator plate are used to regulate the engine timing. Proceed as follows:

a. Be sure the high tension cable and other wires have been properly replaced—or installed—that there is no rubbing of these wires with any moving part. Also make certain there are no breaks or pin holes in the high tension cable right up to where it is attached to the coil. If any breaks are found, repair temporarily with electrician's insulating tape. If break is too large—replace the complete coil assembly with the exact replacement part.

b. Refer to Fig. 251, which shows a standard breaker magneto assembly mounted in the

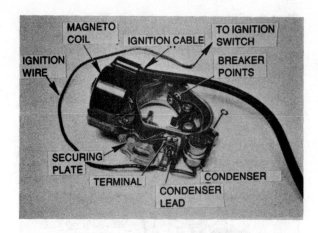

FIGURE 253 Typical Breaker Points Magneto

FIGURE 254 Breaker-less Magneto

FIGURE 255 Solid State Breaker-less Magneto

engine. Note the elongated holes in the stator plate which are used to secure the assembly onto the engine.

c. From the position of looking down onto the magneto assembly, loosen the stator plate mounting bolts and rotate the entire assembly clockwise. Rotate as far as it will turn, then tighten the bolts.

d. This completes the timing procedure and sets the timing, 22° BTC, in this example.

Step 2—*Timing, using marks stamped on flywheel.* Some engines, such as the Kohler engine, are equipped with timing marks which are stamped onto the flywheel magneto. A timing sight hole located in the blower housing (or the bearing plate) permits direct viewing of the timing mark. This sight hole has a snap button which covers the hole when not in use. Refer to Fig. 256, and proceed as follows:

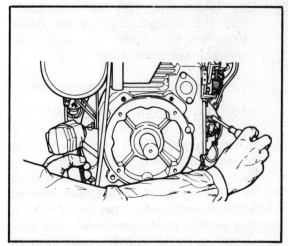

FIGURE 256 Engine Timing Using Timing Light and Sight Hole

a. Pry the snap button loose with a screwdriver, and remove the button.

b. A "T" mark on the flywheel indicates top dead center, while an "S" or "SP" indicates spark or spark-plug point. Spark-plug point is 20° before top center.

c. Refer back to Step 2a and note the operation of the externally located breaker points. Be sure to remove the ignition cable from the spark plug.

d. Rotate the flywheel by hand (or starter) slowly, in a clockwise direction when viewed from flywheel end, until the "S" or "SP" mark appears in the center of the sight hole.

e. The points should just begin to open at this point. If they do not, loosen the lockdown screw on the points set and rotate set until they just begin to open. Then tighten lock screw.

f. Continue rotating the flywheel until the points are at the maximum opening.

g. Measure the gap with a 0.020 inch feeler gage.

h. If the gap is not 0.020, reset or adjust as per Step 2 after loosening the lockdown screws. Gap tolerance is from 0.018 to 0.022. Check fit as per pg. 222. After proper gap is set replace snap button over sight hole, replace breaker points gasket and cover.

i. Proceed to next ignition step as required.

• *BREAKER CONDENSER TEST AND RE-PLACEMENT.* When previous tests of the ignition system indicate no spark, or no electrical output, the most probable ignition system component (after the points have been checked good, or have been replaced with new points) which may be defective is the breaker points condenser. The condenser can break down internally, in which case the system output voltage will be reduced. If the condenser develops a short circuit there will be no system output voltage. The condenser size must be correct. One indication of improper size or rating is the transfer of metal from one breaker point to the other—hence the pitting and build up on the points. Frequent recurrence of very badly burned points indicate a poor condenser. The condenser can be tested quickly with an Ohmmeter as follows:

Step 1—Either remove the condenser from the engine completely or remove the condenser body securing screw so that the condenser and its ground strap are lifted up away from the magneto breaker assembly as shown in Fig. 257.

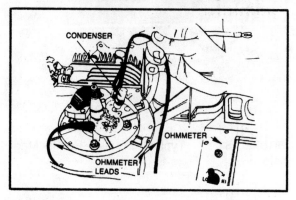

FIGURE 257   Lifting End of Condenser for Test

Step 2—Connect the Ohmmeter to the ground strap on the condenser and to the end of the condenser wire.

Step 3—The Ohmmeter should read a low resistance initially, and then the resistance should rise quickly to a high value. If a low resistance is read continuously the condenser is defective. Replace with the exact replacement part recommended by the engine manufacturer.

• *HIGH TENSION IGNITION CABLE RE-PLACEMENT.* If prior tests have shown that the high tension ignition cable is faulty and is shorting out to ground, this is due to a break in the cable insulation. If inspection of the cable shows the damage to be small it may be repaired by the use of electrician's insulating tape. If the damage is too great, or there is a break in the cable, then the ignition coil assembly, which holds the fixed end of the cable, must be replaced. The cable/coil are usually molded and wired together as one unit and the cable cannot be separately replaced. Replace the coil assembly with the exact replacement part specified by the engine manufacturer.

• *REPLACEMENT OF FLYWHEEL MAG-NETO.* Perform a final inspection on all the parts of the magneto breaker assembly.

Use a dab of lubriplate grease on the cam if new breaker points were installed. Be sure no wires are rubbing against any moving part, and that clearances are adequate. Check all connections and screws to be sure they are secure. Use a new key if necessary and be certain it is installed properly. Replace any cover which fits over the breaker points and then reinstall the flywheel. Replace flywheel starter cup, cupped washer, etc., and then the flywheel nut. Use the exact socket wrench or box wrench size to tighten, then use the flywheel strap wrench to hold the flywheel for final tightening. The flywheel nut must be torqued using a torque wrench as shown in Fig. 258. Refer to the engine manufacturer's specification for the torque value or if unavailable use Table VI as a guide. *REPLACE ALL SHROUDS, AIR VANE*

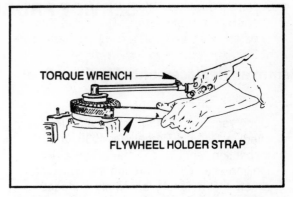

FIGURE 258   Torque Tightening Flywheel Nut

DEFLECTORS, ETC., THE STARTER AND ANY OTHER PARTS WHICH WERE REMOVED. Reinstall the spark plug and connect the ignition cable to spark plug.

TABLE VI

TABULATION OF TYPICAL FLYWHEEL NUT TORQUE VALUES

Engine Manufacturer and Model Number	Torque Value in Inch Pounds
Jacobsen Model 321	300–360
Kohler Model K91 K141, 161, 181 K241, 301, 321	480–660 600–720 720–840
Clinton Model A300, 429, 431, V100, V3100, 406, H3100, and all 4 stroke aluminum vertical and horizontal shaft, 200, A200, VS200, VS400, A400, 500, 501 1600, 1800, 2500, 2790, 414, 416, 418, 420, 422	375–400 1200–1440
All Others	400–450

SERVICE PROCEDURE 4035

Removal, Replacement of Ignition Seal

• Upon removal of the flywheel magneto as cited in Service Procedure **4030,** a close inspection should be made for a leaky ignition seal. Oil film around the crankshaft or on the magneto assembly surfaces is evidence of a leaky seal. The ignition seal will require replacement since a leaky seal will completely destroy the ignition system. Carefully measure and record the amount by which the oil seal protrudes above the block or plate. Proceed to Step 1.

• On many engines the ignition oil seal may be removed and replaced with the crankshaft in place. (If upon inspection of the seal on the crankshaft it is evident that this may be very difficult due to encumbrances or the need to remove the seal from the crankshaft side then it is recommended this task be passed on to the Authorized Factory Service Dealer.) This can be done with the use of an oil seal puller which is shown in Fig. 259. As can be seen, the puller has hooks on the end of a shaft. These hooks are forced down into the seal as shown in Fig. 260, and the seal pulled up and out.

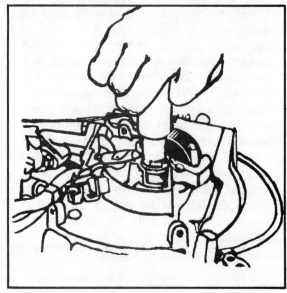

FIGURE 260   Ignition Oil Seal Removal

Step 1—Thoroughly clean the area around the seal seat and procure an exact replacement seal. To replace the oil seal note that the seal usually has a heavy lip that goes down toward the crankcase. Fig. 261 shows some typical oil seals and the heavy lip is on the lower or bottom side of these seals. The seal "B" also has a second lip which is up near the top side or face of the seal. This lip is smaller, however, and care must be taken not to install this seal upside down. If this is done the seal may leak. Fig. 262 shows an oil seal with a small amount of gasket sealer placed around its rim or circumference. This is done to compensate for any out of roundness which the seal may have. After the seal is inserted the rim of the seal is wiped off with a clean cloth which seals off any out of round condition on the base of the block or plate. If the new seal has a coating of neoprene around the outside then the gasket sealer is unnecessary since the neoprene will compensate for any out of roundness. Proceed to Step 2.

FIGURE 259   Oil Seal Puller

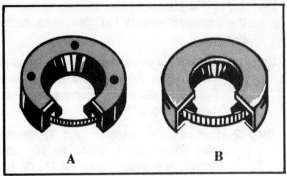

FIGURE 261   Typical Oil Seals

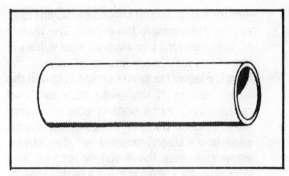

FIGURE 263   Oil Seal Driver

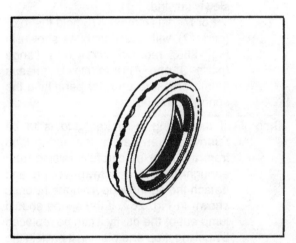

FIGURE 262   Oil Seal with Gasket Sealer

Step 2—To install the new seal it is necessary to use an oil seal driver as shown in Fig. 263. This is a piece of tubing which fits over the crankshaft end and is used to press the seal down into position.

NOTE:

Be careful to insert seal to the same depth as the old seal. Use the information taken on the amount the old seal protruded above the block and make certain new seal protrudes the same amount. Drive the new seal down in small steps until desired measurement is obtained.

Step 3—Refer to Service Procedure **4030** for reinstallation of ignition components, flywheel, etc.

SERVICE PROCEDURE   4040

Retractable Rope Pull Starter, Unit Removal, and Repair

• Gasoline powered lawn mowers use a variety of rope pull starters. This service procedure will cover retractable (also called recoil) rope pull starters which are used on a number of mower engines such as Kohler, Clinton and Jacobsen. A general procedure is given for removal of the starter assembly from the engine. Specific information is then given which discusses releasing the rewind spring tension, replacement of the rope and rope retracting (rewind) spring, and finally reinstallation of the assembly on the engine. Shut engine down, remove ignition cable from spark plug and proceed.

• *REMOVAL OF STARTER UNIT FROM ENGINE.* Different rope pull starter manufacturers use different types of fasteners to secure the starter to the engine. The variety of fastener types is evident from different manufacturer's engines. In all cases the important point to keep in mind is to use the exact tool to fit the given fasteners and remove only those bolts or screws necessary to permit lifting off the complete starter assembly. Upon removal of the starter assembly place the mounting screws in a container for safekeeping. Take the assembly to a work bench or table.

```
┌──────────── CAUTION ────────────┐
│   ALWAYS WEAR SAFETY GLASSES     │
│   WHEN WORKING ON ANY RECOIL     │
│   STARTER. USE CARE NOT TO       │
│   JOLT OR JAR THE STARTER        │
│   SINCE THE RECOIL SPRING MAY    │
│   BE DISLODGED FROM ITS CON-     │
│   TAINER                         │
│   AND CAUSE INJURY.              │
└──────────── CAUTION ────────────┘
```

- *RELEASING THE REWIND SPRING TENSION.* The spring tension must be released before any repair or work can be done on the starter.

For Fairbanks Morse starters shown in the exploded view of Fig. 264, proceed as follows:

Step 1—Hold the washer (part 7) in position with your thumb, then remove the retainer ring (part 6) with a screwdriver as shown in Fig. 265.

Step 2—Remove the washer (part 7), spring (part 8) washers (parts 9, 10) and friction shoe assembly (parts 11, 12, 13, 14).

Step 3—For Model Number 425 remove the rope pull handle to permit the spring (rotor) to slowly unwind.

For Model Number 475 hold the rotor (part 17) with the thumb as shown in Fig. 266, remove screws in flange (parts 3 and 4) and slowly release thumb pressure on rotor permitting the spring to unwind.

Step 4—If the spring is broken and is to be removed, then detach the spring loop from the rotor by carefully raising rotor enough to get a screwdriver in and detach the spring loop from the rotor as shown in Fig. 267. If the spring should jump out of the cavity it can be replaced by coiling it up and placing it back in the cavity.

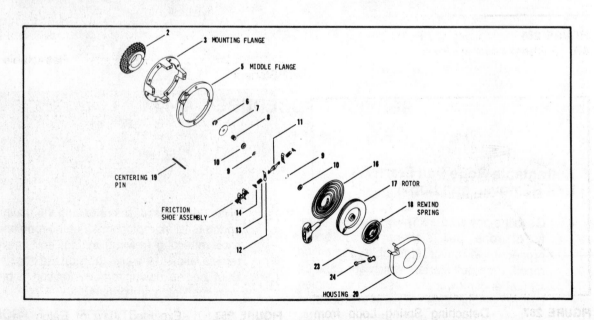

FIGURE 264   Exploded View of Fairbanks Morse Retractable Rope Pull Starter

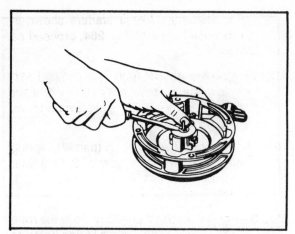

FIGURE 265   Removal of Retainer Ring (Part 6)

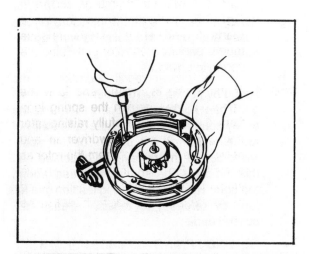

FIGURE 266   Unwinding Rotor on Model No. 475 (Fairbanks Morse)

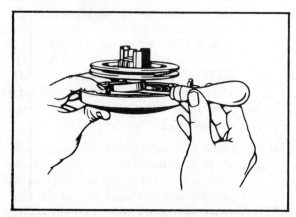

FIGURE 267   Detaching Spring Loop from Rotor

For Eaton retractable starters shown in Figs. 268 and 269, proceed as follows to release the spring tension:

Step 1—Hold the starter with the pulley side up as shown in Fig. 270.

Step 2—Hold the thumb over the pulley so as to keep pulley from rotating, then pull rope out about 12 inches. Hook the rope where it passes through the starter housing, with a screwdriver.

Step 3—Pull the rest of the rope out and while lightly holding the pulley with the thumb, turn the pulley clockwise with the rope, thereby permitting the spring to uncoil. Continue until there is no more tension on the rope.

• *ROPE REPLACEMENT.* To replace a broken or frayed rope for the *the Fairbanks Morse starter* proceed as follows:

FIGURE 268   Typical Eaton Retractable Starters

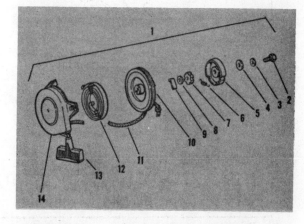

FIGURE 269   Exploded View of Eaton Retractable Starter (Reassembly Sequence)

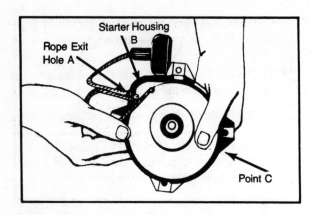

FIGURE 270   Tension Release on Eaton Starter

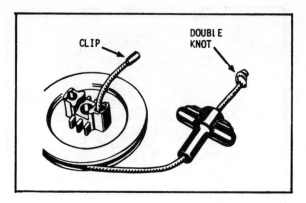

FIGURE 271   Rope Replacement

Step 1—Refer to Fig. 271. Install the new rope into the rotor through the rotor hole with the metal clip end first. Be sure rope is wound in the proper direction (counter clockwise as shown).

Step 2—Replace handle and washer and tie double knot in end of rope.

To replace a broken or frayed rope for the *Eaton starter* proceed as follows:

a. Refer to Fig. 270 and insert a new rope through the hole in the pulley. Tie a double knot in rope end, on top of pulley.

b. Work other end of rope through the pulley and the rope exit hole (A). The rope exit hole is in the starter housing (B).

c. Replace the handle grip and tie a double knot at end of rope.

d. Pull the rope through exit hole (A).

e. Grasp the rope at point (C) at the pulley, where the rope enters the pulley, and holding the rope securely wind the spring. Wind the spring by pulling the pulley around, counterclockwise, about four revolutions. Hold the pulley with your thumb while winding.

f. Remove any slack from the rope by pulling rope out of the exit hole.

g. Slowly release the thumb pressure on the pulley and permit the pulley to wind up the rope.

h. Check for proper recoil operation of the spring by pulling the rope out as far as it will go. Then holding the rope out, rotate the pulley counterclockwise. If the pulley will not rotate ¼ turn of the pulley counter clockwise, before the pulley has bottomed out, this indicates the spring has been wound too tight. Release the spring tension one full turn by referring to tension release discussed earlier.

● *REWIND SPRING REPLACEMENT.*

Fairbanks Morse Starter

Step 1—Refer to Fig. 272. Note position of inner and outer spring loops and begin to remove old spring by starting at the center of the housing. Pull innermost loop out first with a pair of pliers. HOLD THE SPRING END FIRMLY; DO NOT PERMIT SPRING TO WHIP ABOUT. Hold down the outer coils with your thumb so that spring does not erupt suddenly. see numbers on Fig. 264.

Step 2—Place new spring in same position, with inner and outer spring loops, as old spring. Note the proper position has the outside loop engaged around the centering pin (part 19). The new spring should be supplied in a spring holder and after positioning the spring holder

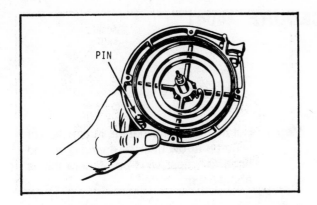

FIGURE 272 Position of Inner and Outer Spring Loops

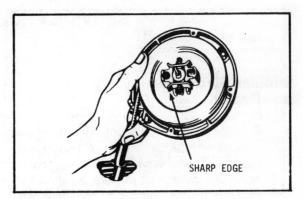

FIGURE 273 Pre-tensioning the Starter

over the pin and cavity, push spring down into cavity. Spray or brush a few drops of SAE 20 oil over the spring. Light grease should be placed on the cover shaft.

Step 3—For Model Number 475 place the rotor and cord into the cover. Hook the inside loop of spring to the rotor as shown in Fig. 267. For Model Number 425 be sure the rope is fully wound around the rotor before placing into cover.

Step 4—Replace washers (parts 9, 10), friction shoe assembly (parts 11, 12, 13, 14), spring, washer and rotating ring (parts 8, 7, 6). Refer to Fig. 264.

Step 5—The starter must be pre-tensioned. Refer to Fig. 273. For Model Number 475 wind the rotor, with the aid of the cord, four additional turns in the same direction the cord is wound. For Model Number 425 wind the rotor five additional turns.

Step 6—Replace the flanges (parts 5, 3) for both Models Number 475 and 425 by holding starter as shown in Fig. 266 and replacing fastening screws.

Eaton Starter

Step 1—Make certain the spring tension has been relaxed as covered previously.

Step 2—Note the manner in which the old spring is installed with regard to the winding

direction. Use caution in removing old spring. Hold spring end firmly with pliers or vise-grips. Be careful the spring does not jump out of its cavity.

Step 3—Leave the new spring in its spring holder and use care to position the new spring exactly as the old one was positioned.

Step 4—Spray or brush a few drops of oil over the new spring and be sure all parts are clean.

Step 5—Reassemble the starter in the sequence shown in Fig. 269.

• *REINSTALLATION ON ENGINE.* The Eaton starter assembly is reinstalled on the engine by the simple reversal of the removal steps. For Fairbanks Morse, the starter must be properly centered on the crankshaft or it will be damaged. Proceed as follows:

Step 1—Refer to Fig. 264. Pull the centering pin (part 19) slightly out, about 1/8 inch.

Step 2—Place the starter assembly on the engine making sure the centering pin engages in the center hole of crankshaft, then press down into position.

Step 3—Hold the starter in this position and attach securing screws or nuts and lock washers. Tighten securely.

Step 4—Reconnect the ignition cable to spark plug.

SERVICE PROCEDURE   4041

Impulse Starter, Removal and Repair

- As in the case of rope pull starters, there are a variety of impulse, or wind up, lawnmower starters. These service procedures cover some typical impulse starters, as shown in Figs. 274, 275. Shut the engine down, remove ignition cable from spark plug and proceed to Step 2.

- In some cases the impulse starter is believed to be inoperative when actually it may not be wound properly. The throttle control lever must position the choke to full choke before wind up of the starter, and until after the engine starts. Refer to Figs. 276, 277 for starter activation as follows:

Step 1—Rotate the starter handle in direction of wind-up arrow until the spring is wound tight. On most wind-up starters this is about 6 turns.

Step 2—For button release starters as shown, push down on the starter handle holding the handle hard against the stop until the starter releases. Sometimes a momentary hesitation will occur between pushing down on the starter and the engine cranking. The same procedure holds for impulse starters which have a mechanical lever or lug release

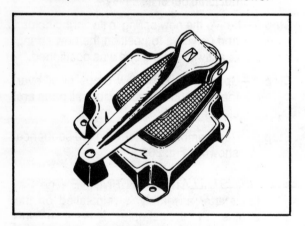

FIGURE 274   Typical Impulse Starter

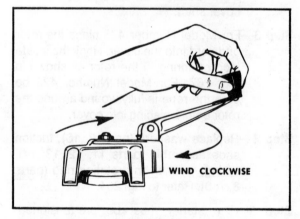

FIGURE 276   Starter Wind-up

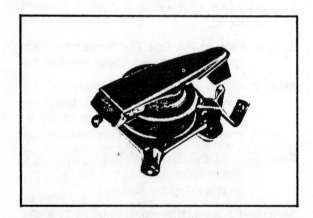

FIGURE 275   Typical Impulse Starter

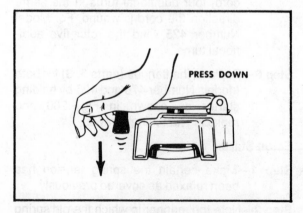

FIGURE 277   Starter Release

rather than a center push down button release. This hesitation is from the piston traveling through its compression stroke.

— CAUTION —

ALWAYS WEAR SAFETY GLASSES WHEN WORKING ON ANY RECOIL STARTER. USE CARE NOT TO JOLT OR DISLODGE SPRING. SPRING CAN JUMP OUT OF ITS CONTAINER AND CAUSE INJURY.

— CAUTION —

- REMOVAL OF STARTER UNIT FROM ENGINE. There is no one procedure which applies to a removal sequence for all of the various impulse starters from the mower engine. The shroud, baffle, air deflector, etc. must be carefully examined to determine the location and type fasteners securing the starter to the engine. Be sure to use the exact tool for the given fastener whether they are bolts or screws. Also be certain to release the starter spring before removal from engine. Should the spring not release, tap the housing lightly before removal. If the spring still does not release very carefully remove the assembly from the engine. Remove the screws holding the bottom cover to the starter housing (part 10). See Fig. 278. Place assembly legs down, on clean surface, and holding the starter at arm's length with hand on top of housing, lightly tap the assembly. This should dislodge the bound-up spring which will be evident by the noise the spring makes as it releases.

- INSPECTION PROCEDURE. The starter is available to be inspected for defective parts at this point. Remove the ratchet with a 3/8 inch Allen wrench. Do not remove the power spring from power spring cup since this is serviced and replaced as a spring cup assembly. Mark the position of the

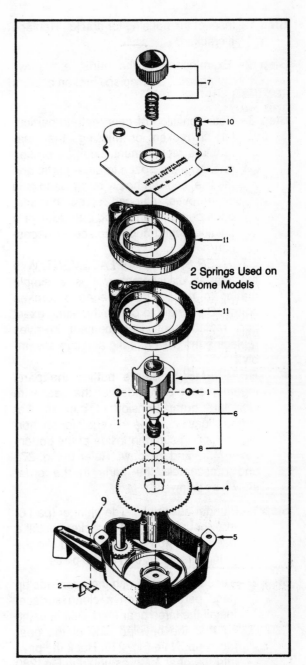

2 Springs Used on Some Models

FIGURE 278   Exploded View of Wind-up Impulse Starter

spring and cup assembly for later replacement in same relationship. Remove the bottom cover assembly from the internal assembly and the plunger assembly from power spring and cup. Be extra careful not to force the spring out of the cup. Clean all parts in a cleaning solvent and perform the following visual inspection.

Step 1—Inspect the housing for cracks. Replace if cracked.

Step 2—Examine all of the parts, e.g. pawl, spring, etc., and replace broken or worn parts.

Step 3—Note condition of gear teeth in particular; if broken or missing, the gear requires replacement. Look for broken ratchet, parts frozen to other parts, and the like. Look for heavy burrs, deep cuts or grooves, stripped screw threads, damaged uneven spring cup sides or a bulging spring cup. Replace as necessary.

• *STARTER SPRING REPLACEMENT.* With the starter disassembled it is a simple matter to replace a damaged or broken spring. Be certain to obtain the same exact part replacement. Replacement involves replacing the power spring and cup assembly.

• *REASSEMBLY.* Before putting the parts together coat both sides of the gear with lubriplate grease, inside of cup so that spring slides easily, handle shaft for anchoring the spring cup, the inside of the bottom cover plate and the pawl. Refer to Fig. 278 and reassemble the parts in the order shown as follows:

Step 1—Plunger assembly, slide plunger (part 6) into the bushing and sprocket, install a ball (part 1) on each side, and slip the plunger retainer (part 8) into place.

Step 2—Place the housing with the open side up (legs part 5 up) on a clean surface. Install the large gear (part 4) in housing so that the beveled side of the gear faces up. Note Fig. 279. Hook the pawl into the gear teeth as noted in Fig. 280 and push the gear sideways to allow the gear to move into position. Now hold the large gear in position so that it is in the center of the housing and in alignment with the small gear on the handle shaft. The large and small gear teeth should be meshed together. Install the plunger assembly through the hole in the large gear, down into the housing. The end of the plunger (the flat release end) should protrude through the starter housing top.

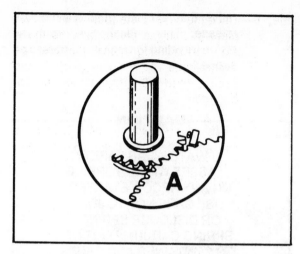

FIGURE 279   Pawl Hooked into Gear Teeth

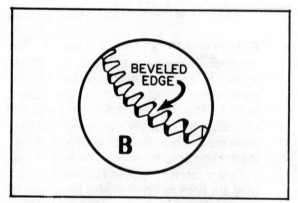

FIGURE 280   Large Gear Installation

Step 3—Install the power spring and cup assembly (part 11) with the open side of the cup facing up as shown. The cup assembly slides over the plunger (part 6), and the outside cup boss hole goes over and down on the handle shaft. For those starters which have two springs install the second spring in the same manner as noted in Fig. 278.

Step 4—Using care so as not to push the spring out of the spring cup, hook the center part of the spring end over the plunger assembly using a screwdriver or long needle nose pliers.

Step 5—When the spring end is hooked over the plunger assembly, carefully push the cup assembly spring down into the large drive gear (part 4).

Step 6—Place the cover plate (part 3) with cover plate bushing in place (be sure thrust side of bushing faces towards the inside) onto the housing (part 5) and secure (part 10). Tighten to 20 inch pounds.

Step 7—Place plunger spring (part 7) into the plunger hole. This spring is very important. Omitting the spring will result in an inoperative starter.

Step 8—Screw ratchet (part 7) into the plunger assembly. Tighten with a 3/8 inch Allen wrench.

- *REINSTALLATION ON ENGINE.* Reinstallation of the starter assembly is relatively simple and involves a reverse order of Step 3. Be sure to replace all screws or bolts and shroud, blower housing or other parts which may have been removed to remove the starter assembly. Reconnect the ignition cable.

SERVICE PROCEDURE   4042

Electric Starter Motor Overhaul

- The electric starting motor used on lawn mower engines so equipped generally uses a Bendix type drive to engage and disengage from the ring gear on the engine. A typical drive gear arrangement with the starter motor is shown in Fig. 281. When the start circuit is energized the starter motor armature begins to rotate and the drive pinion moves along a splined sleeve to mesh with the engine ring gear. The starter pinion gear is then driven by the starter motor and in turn cranks the engine. This position is maintained until the engine fires. The engine then builds up to a speed where the engine ring gear overrides the starter armature. As a result the ring gear forces the pinion out of engagement and the pinion goes back to a retracted position. When the starting motor armature comes to a stop a small spring holds the pinion in the retracted position.
- The following precautions should be taken with electric starter motors:

Step 1—Limit the time the engine is cranked by the starter motor to a maximum of one minute. Always permit this to be followed by a cooling period of about one minute.

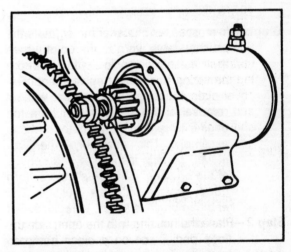

FIGURE 281   Typical Electric Starter Motor Drive Gear

Step 2—In the event of a "false start" and the engine approaches enough speed to disengage the starter, but does not continue to run, always have the engine come to a full stop before restarting. If this is not done the engine ring gear and starter gears can clash together and either lock or break off gear teeth. Shut the engine down, and remove the battery cable before servicing the starter.

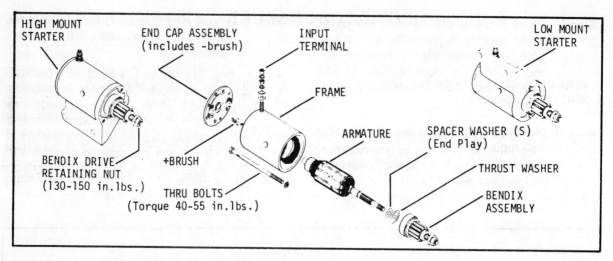

FIGURE 282   Exploded View Typical Electric Starter Motor

● Refer to Fig. 282 for an exploded view of a typical starter motor and proceed to service the starter motor brushes as follows:

Step 1—In most cases of starter motor faults the brushes have worn to the point where they are no longer in contact with the armature and need replacement. Replacement can usually be made without removal of the starter motor from the engine by removing the motor end cap assembly. This assembly when removed permits inspection and serivce of the brushes. If the brushes are worn down, replace with new brushes. The end cap brush is simple to replace. But the positive brush is fastened to the field winding and is somewhat harder to replace since it is difficult to reach. If this brush becomes too difficult to reach due to interference with the engine proper, remove the starter motor. Then work on the starter motor on a bench.

NOTE:

Some brushes require removal of a fastening rivet and use of a new rivet to hold the new brush. In these cases the motor must be removed.

Step 2—When the brushes are removed the commutator is accessible to be examined or cleaned. If the commutator is grooved or dirty it should be cleaned by the use of a commutator stone or fine sandpaper. Hold the stone on the commutator and rotate the armature by hand until the grooves or ridges are cleaned up.

Step 3—When reinstalling the end cap put a light coat of SAE 10 oil on the bushing and end of the armature shaft. Be certain there is no excess oil which can spatter from these parts. Refer to Fig. 282. Hold the positive brush spring away from the brush with a pair of needle nose pliers and guide the end cap into place. Release the brush spring when brushes seat on the commutator. Secure the cap to frame by means of the through bolts. Tighten to 40-55 inch pounds of torque.

Step 4—Service of the Bendix drive is limited to inspection for damage and replacement as needed by a new assembly. Inspect the drive for damaged teeth or a drive gear that is jammed on the shaft and cannot be moved.

Step 5—When reinstalling the starter motor be certain to use any special mounting bolts. These special bolts provide a proper alignment of the starter drive gear with the engine ring gear. Reconnect battery cable and spark plug.

SERVICE PROCEDURE   4050

Engine RPM Settings for Idle and Fast Speeds

1. A properly performed engine speed or RPM adjustment requires accurate RPM reading for setting both the IDLE and POWER carburetor adjustment screws (or needles). The service procedure for set up and reading of engine RPM is given in Service Procedure **4010**.

2. Service Procedure **4050** Table VII provides a tabulation of typical engine manufacturer's IDLE and POWER (or fast) RPM settings to be used in conjunction with fine tuning of the engine speed. For engines not listed in Table VII contact the engine manufacturer or Authorized Factory Service Dealer for this information

TABLE VII

TABULATION OF VARIOUS ENGINE IDLE AND POWER RPM SETTINGS

Engine Manufacturer and Model Number	RPM Settings IDLE	POWER (or fast)
Clinton	set IDLE screw based on carburetor used see Table	3600 (maximum)
Kohler		
K91	1000	4000 (maximum)
K141, K161, K181	1000	3600 (maximum)
K241, K301, K321	1000	3600 (maximum)
Jacobsen 321 Engine		
18″ Pacer (52C18, 42D18, 42E18)	1300 to 1600	up to 3500
18″ Pacer (11814)	1500 to 1800	up to 3600
21″ Lawn Queen (2C21, 2D21, 2E21)	1300 to 1600	up to 3500
21″ Lawn Queen (12113)	1500 to 1750	up to 3600
21″ Manor (28F21, 28G21)	1300 to 1600	up to 3500
21″ Manor (22114, 32121–7B1)	1500 to 1750	3200 to 3600
22″ Greensmower (9A22, 9B22)	1500 to 1800	up to 3400
22″ Greensmower (62203, 62208)	1500 to 1750	3400 to 3800
24″ Estate (8A24, 8B24)	1300 to 1600	up to 3800
26″ Estate (8A26, 8B26, 8C26, 8D26)	1300 to 1600	up to 3800
26″ Estate RR (22601, 22605–7B1)	1500 to 1750	3400 to 3800
26″ Estate FR (22611, 22615–7B1)	1500 to 1750	3400 to 3800
26″ Lawn King (12A26, 12B26)	1300 to 1600	up to 3200
26″ Lawn King (12601)	1500 to 1800	up to 3800
18″ Turbo-Cut (3418, 34B18, 34C18, 34D18)	1300 to 1800	3400 to 3500

Engine Manufacturer and Model Number	RPM Settings IDLE	POWER (or fast)
18″ Turbo-Cut (7518, 75A18, 75B18)	1500 to 2000	3400 to 3500
18″ Turbo-Cone (117–18)	1500 to 1600	3200 to 3400
18″ Turbo-Vac (31819)	1500 to 1800	3200 to 3400
18″ Turbo-Vac (31819)	2400 to 2600	3200 to 3400
18″ 4-Blade Rotary (31809)	2400 to 2600	3200 to 3400
20″ Commercial (32028)	2400 to 2600	3200 to 3400
20″ Commercial Rotary (32028)	1500 to 1600	3200 to 3400
20″ Robust (32031)	1500 to 1700	3200 to 3400
20″ Scepter (8020, 80A20)	1500 to 1800	3000 to 3200
21″ 4-Blade Rotary (32114)	2400 to 2600	3200 to 3400
21″ 4-Blade Rotary SP (42114, 42118, 42119)	2400 to 2600	3200 to 3400
21″ Turbo-Cut (3921, 39B21, 39C21)	1500 to 1800	3200 to 3400
21″ Turbo Cone (119–21)	1500 to 1600	3200 to 3400
21″ Turbo-Cut (3521, 35C21, 35D21, 35E21, 35F21)	1500 to 1800	up to 3500
21″ Turbo Cone (121–21)	1500 to 1600	3200 to 3400
22″ Scepter (80A22)	1500 to 1800	3000 to 3200
24″ Rotary SP (40A24)	1900 to 2100	up to 3000
Tecumseh		
Series SLV, LV, LAV, LCV, LVR, LAVR, LAVT, LAVH, LAVTC, LAVTR	1800	3200 to 3600
Series VA, VC, VH, VX, VCX, VT	1800	3200 to 3600
Series H, HR, HA, HTB, HTC, HRT, HT, HH	1800	3200 to 3600
Briggs and Stratton		
Series A, WI, 5S, 6S 6HFB, 6HS	1750	3200
Series 5, 6, 8, N	1750	3600
Series B	1750	2400
Series Z, ZZ	1750	3100

SERVICE PROCEDURE 4060

Cylinder Head Gasket Replacement: Carbon Removal and Head Bolt Tightening

- This procedure outlines the steps to be taken for removal of the engine cylinder head, and replacement of the cylinder head gasket. Fault Indication **2500** describes the test which determines a leaky head gasket. Generally it should not be necessary to have to remove many parts of the engine to gain access to the cylinder head. The procedure which follows applies to both 2- and 4-cycle engines.

- Refer to Fig. 284 for a sketch and an exploded view of engine components which include the cylinder head gasket. Proceed as follows:

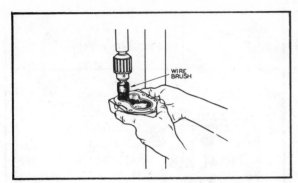

FIGURE 283 Cylinder Head Clean-out of Carbon

Step 1—Clean off all parts of the engine. Be sure to clean out the air fins around the cylinder and on the cylinder head.

Step 2—Disconnect ignition cable and remove the spark plug (see Service Procedure **4004**).

Step 3—Remove, as required, any shroud, air deflector, blower cover or blower housing to fully expose the cylinder head. There must be no encumbrances around the cylinder head which will keep the head from separating from the cylinder.

Step 4—Shut off the fuel valve or line to keep fuel from flowing to the carburetor. If necessary, remove carburetor (refer to Service Procedure **4011**).

Step 5—Remove the cylinder head bolts. Use the exact size box wrench or a socket wrench. See Fig. 297 for typical wrenches which are used for this task. Place the bolts into a can or container for safekeeping.

Step 6—Remove the head. Should the head stick and not come loose easily and all fasteners and encumbrances have been removed, use a flat thin scrapper to pry or separate the head from the cylinder. Work the scrapper all around the cylinder being careful not to burr or mar the surfaces or edges.

Step 7—When the head is removed, clean off the flat machined surfaces and the cylinder. Clean off all of the old gasket and sealer such that a clean smooth surface is obtained. Refer to Fig. 283. Scrape all carbon deposits and foreign matter from both the head and the top of the cylinder until they are clean. Use a wood spatula to scrape the carbon from the top of the piston.

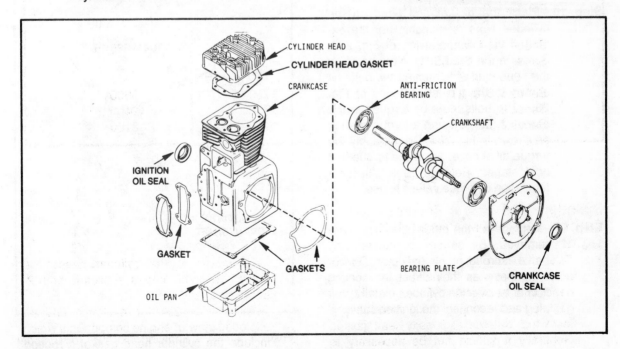

FIGURE 284 Exploded View Typical Major Parts of 4-Cycle Engine

Step 8—Examine the head for stripped threads in the spark-plug hole and for broken cooling fins. If either of these conditions are found replace the head with a new head. If the head is slightly warped this can be corrected. Check the head warpage by placing the head on a thick flat piece of glass such as is used for desk top. If the head is warped, the unevenness will be evident by inspection. Place an 8 x 10 sheet of emery cloth on this flat surface, rough side up. Place the head machine side down on the emery cloth and move the head about in a figure-8 movement on the emery. This will cut away the warpage. Work the head over the emery until the head is flat. Keep checking the head by placing it on the glass plate until there is no sign of unevenness on the head. Be sure to clean away all the emery and dirt from the head before reinstalling on cylinder.

Step 9—Obtain an exact cylinder head gasket replacement. Be sure the head gasket matches the head contour and has the same cutouts.

Step 10—Place the gasket and head on the cylinder. Note relationship per the exploded view line drawing of Fig. 284. Screw in the head bolts. Note carefully the sequence of tightening the bolts as shown in the typical sketches of Fig. 285. The bolts must be torqued in the same sequence using a torque wrench as shown in Fig. 286. Do not apply the torque all at once. Apply the torque in 3 or 4 stages. Refer to Table VIII for a tabulation of torque values to use.

Step 11—Replace all the other parts that were removed. Be certain to reinstall any baffle, shroud or air deflector. Do not omit these as they direct the cooling flow of air over the cylinder. Install spark plug and reconnect the ignition cable.

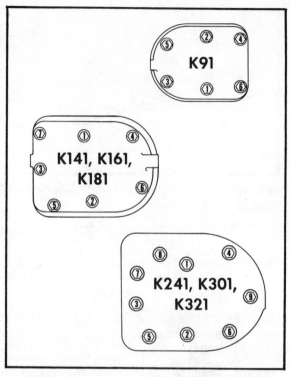

FIGURE 285   Typical Bolt Tightening Sequences for Cylinder heads (Kohler Engine Models Noted)

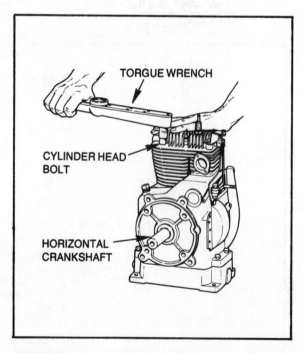

FIGURE 286   Applying Torque to Cylinder Head Bolts

TABLE VIII

CYLINDER HEAD BOLT TORQUE FOR SELECTED LAWN MOWER ENGINES

Engine Manufacturer and Model Number	Torque in Foot Pounds
Jacobsen Model 321 engine	12½–15
Kohler	12½–15
K91	15½
K141, 161, 181	15–20
K241, 301, 321	25–30
Clinton	
All 4 stroke aluminum vertical and horizontal shaft	10–12½
V100, V3100, 406, 407, 417, H3100	18 3/4–21
others	15½–18 1/3

SERVICE PROCEDURE   4070

Cylinder Crankcase Gasket Replacement

1. Replacement of an engine cylinder crankcase gasket is a major task requiring nearly complete disassembly of the engine. This is due to the fact that in many engines the crankshaft must be removed since the crankcase plate contains the housing for one of the crankshaft main bearings. The crankcase plate and bearing are press fit together, and these in turn, onto the crankshaft. Therefore, the crankcase plate is not removed from the engine cylinder as a separate plate by the simple removal of the crankcase plate's securing bolts.

2. Refer to Figs. 284, 288, 289, 290, 291 and 292, which show some typical engines in various stages of disassembly. Note in Fig. 288 that the carburetor assembly and the reed valve assembly must first be removed to expose the connecting rod assembly and its attaching bolts. This connecting rod assembly is then removed before the crankshaft can be removed. It is evident from these figures that a considerable number of engine parts must be removed before the crankshaft is in turn removed. In addition in some cases the use of a special tool is required,

as shown in Figs. 291 and 292.

3. On the basis of the foregoing it is not recommended that most lawn mower owners attempt to replace a leaky engine cylinder crankcase or bearing plate gasket. This major task should be left to the more experienced do-it-yourself mechanic or taken to the engine manufacturer's Authorized Service Dealer.

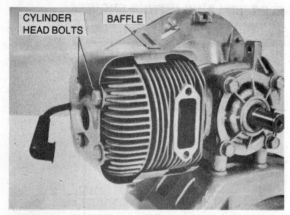

FIGURE 287   Engine Removed from Mower, Preparatory to Removal of Air Baffle and Cylinder Head

FIGURE 290   Crankshaft (with Head Removed)

FIGURE 288   Preparing to Remove Connecting Rod Cap (A), Bolts (C), and Connecting Rod/Piston (B)

FIGURE 291   Removal of Seal (Ignition Seal) Using Arbor Press

FIGURE 289   Removal of Engine Crankcase Head (and Crankshaft)

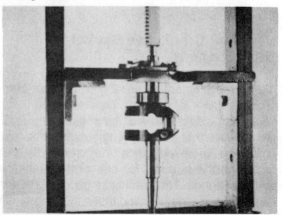

FIGURE 292   Removal of Crankcase Head Bearing (and Crankshaft Oil Seal) Using Arbor Press

SERVICE PROCEDURE   4100

FIGURE 293   Removal of Flywheel

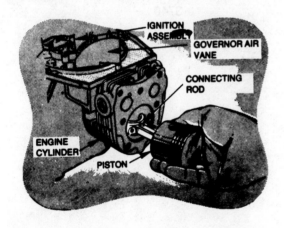

FIGURE 294   Removal of Engine Piston and Connecting Rod (Typical)

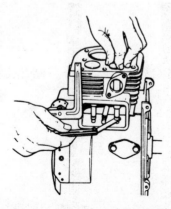

FIGURE 295   Checking/Adjusting Tappet Clearance for 4-Cycle Engine Intake/Exhaust Valves

Major Engine Overhaul for Compression, Cylinder Rings, Valves, Crankshaft, Connecting Rod, or Crankcase Seal

1. Generally most of the **4000** Series Service Procedures can be performed in a relatively simple manner. The Service Procedures involved in a major engine overhaul, or its constituent parts, are complex and usually involve considerable effort and mechanical knowledge. Service Procedures in this series **4100** are in this latter category. For instance to accomplish the task of engine overhaul for correction of poor cylinder compression most of the engine must be disassembled. As illustrated in the Figs. 293, 288, 287, 289, the flywheel, stator assembly, any baffling, the cylinder head, the carburetor, and reed valve assembly must be removed before the connecting rod and piston, Fig. 294, can be removed from the engine block. For 4-cycle engines, removal of the intake/exhaust valves is likewise an involved procedure requiring engine disassembly to the extent shown in Fig. 295.

2. Removal of most engine crankcase seals is also involved. The engine must be torn down to the extent shown in Fig. 289. A special tool, an arbor press, must be used to remove the crankcase seal as shown in Fig. 291. The new seal installation requires use of the arbor press as indicated in Fig. 291. Finally on some engines removal/replacement of the crankcase head seal requires almost complete engine tear down for removal of the crankshaft. The crankshaft is pressed onto the head bearing and seal. Note in Fig. 290 that the crankcase head is removed with the crankshaft attached. The head, bearing and seal must all be pressed off the crankshaft in an arbor press.

3. It is evident from the foregoing that a major overhaul is a formidable task requiring special tools and equipment not usually available in the home workshop. Additionally, it requires considerable time and experience to accomplish this job. It is recommended, therefore, that major overhauls be accomplished by an Authorized Service Dealer.

4. Major overhauls are represented by the following:

Item	Fault Symptom Number
Internal engine damage	1250
Bad compression	2550
Defective piston	2550
Open exhaust valve	2550
Sticky valve	2550
Leaky crankcase seal	2610, 2620, 3150, 3442
Internal mechanical failure	3921
Tappet clearances	3370
Heavy carbon in cylinder	3370
Poor compression	3407
Leaky valve/poor springs	3441
Engine burning oil	3740
Low compression	3740
Intake/exhaust valves leaking	3912

SERVICE PROCEDURE   4200

**Illustrations of Typical
Small Tools Used for
Lawn Mower Repairs**

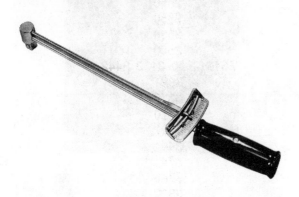

FIGURE 296   Typical Torque Wrench Used
to Torque Spark Plug, Cylinder Head Bolts, etc.

FIGURE 299   Set of Hex Key Wrenches for
Use on Pulleys, Sprockets, and Other Areas
Using Recessed Hex Screws

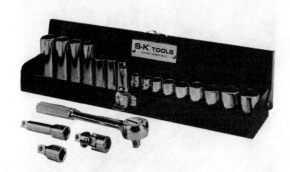

FIGURE 297   Typical   Set   of   Socket
Wrenches, with Ratchet Handle, Universal Drive,
etc.

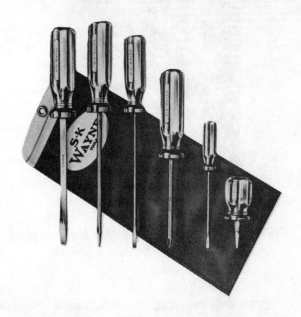

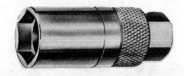

FIGURE 298   Typical Spark-Plug Socket

FIGURE 300   Set of Straight Blade Screwdrivers, Phillips Screwdrivers, and Stubby Screwdriver

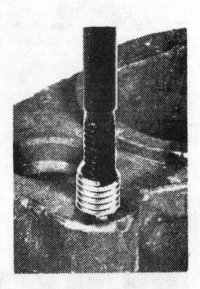

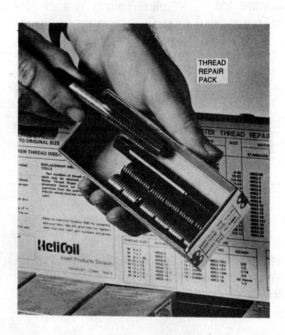

FIGURE 301   Repair Process for Damaged Spark Plug Hole

Spark plug thread repair with the use of "Heli-Coil" inserts and tools is a simple two-step process:

1. The damaged tapped hole is drilled or reamed out and re-threaded with a piloted reamer-tap, assuring perfect alignment and seating of the reinstalled spark plug.
2. The insert is screwed into the "Heli-Coil" tapped thread, returning the hole to its original condition.

FIGURE 302   Typical Volt Ohmmeter

FIGURE 303   Magneto   Coil,   Condenser   Tester

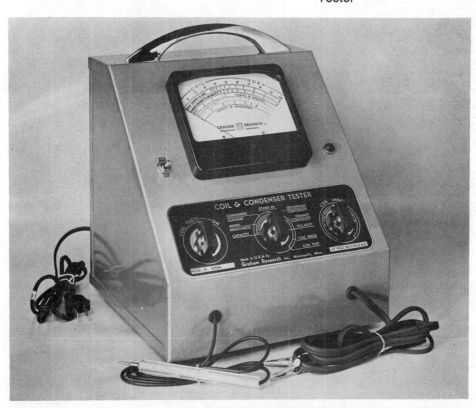

FIGURE 304   Coil and Condenser Tester with Probes and Leads

SERVICE PROCEDURES 5000

Engine Tune-ups and Recommended Periodic Service

- For trouble-free and lasting lawn mower service your mower requires frequent but minor service throughout the cutting season; some major servicing at the end of the season and at longer intervals during the life of the mower. The following is a proposed schedule of maintenance operations.

MAINTENANCE OPERATIONS:	Each use	10 hrs 2 wks	25 hrs 4 wks	50 hrs mid sea.	as req'd.	End sea.	End 2 sea.	End 3 sea.
a. Clean out all dirt, grass clippings, and other debris from the mower top deck.	▓							
b. Check for loose screws, bolts, others parts, especially the blade, mounting bolts/screws.		▓						
c. Oil the wheel bearings and height adjusting linkage with a few drops of oil.		▓						
d. Remove and clean (re-oil as required) the air cleaner element (refer to Fault Symptom **2140**).		▓						
e. Remove, clean and re-gap the spark plug (refer to Service Procedure **4004**).			▓			▓		
f. Change the crankcase oil.			▓					

MAINTENANCE OPERATIONS:	Each use	10 hrs 2 wks	25 hrs 4 wks	50 hrs mid sea	as req'd	End sea.	End 2 sea.	End 3 sea.
g. Remove the air deflector, or engine shroud, and clean all debris, grass, dirt or grease from cylinder cooling fins, flywheel, screens and the like.				■				
h. Adjust carburetor (refer to Service Procedure **4010,** Step 13) and adjust governor speed (refer to Fault Symptom **3450**).					■			
i. Clean out the carburetor, fuel line, pump and fuel tank. Use a commercial cleaner added to fuel tank.					■		■	
j. Test engine compression (refer to Fault Symptom **2500**).						■		
k. Remove, clean or replace oil filter (if there is one on engine).						■		
l. Inspect reed valve on 2-cycle engines (refer to Service Procedure **4012**).						■		
m. Clean carbon from muffler and exhaust parts on 2-cycle engines (refer to Fault Symptom **3700**).						■		
n. Remove flywheel and inspect/ replace ignition breaker points as required (refer to Service Procedure **4030**).								■
o. Adjust governor speed (refer to Fault Symptom **3450**).				■				

SERVICE PROCEDURE   5010

Winter Storage

- If these simple steps are followed you will extend the life of your mower and have trouble-free start-up when required.

Step 1—Drain all fuel from the fuel tank.

Step 2—Run the engine so as to use up any remaining fuel in the tank and fuel lines. Operate the CHOKE lever to help drain the carburetor as the engine is sputtering.

Step 3—Remove the carburetor bowl, if there is one, and clean the bowl thoroughly with carburetor cleaner. Replace when dry.

Step 4—Remove, clean and regap the spark plug (refer to Service Procedure **4004**).

Step 5—Squirt about 10 to 15 drops of lubricating oil inside the cylinder through the spark-plug hole. Crank over the engine several times to evenly distribute the oil inside the engine. Replace spark plug and torque plug (refer to Service Procedure **4004**).

Step 6—Drain the crankcase oil (for 4-cycle engines) and replace with fresh oil.

Step 7—Remove ignition cable from spark plug, and remove (rotary) blade (or check reel for sharpness), sharpen as required (refer to Service Procedure **4002** or **4003**, as applicable). Replace sharpened blade and spray a light coat of oil over blade(s) and any other non-painted exposed metal surface.

Step 8—Disconnect the charger unit in mowers equipped with a power pack.

Short Period Storage

- Whenever the lawn mower has to stand for one to several weeks without running the engine, gum can form in the fuel tank and carburetor. This gum can clog the carburetor and cause operating difficulty. To avoid this difficulty, keep the fuel tank full after each use.

SERVICE PROCEDURES 6000

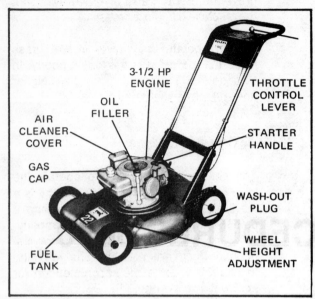

FIGURE 305 Mower Nomenclature

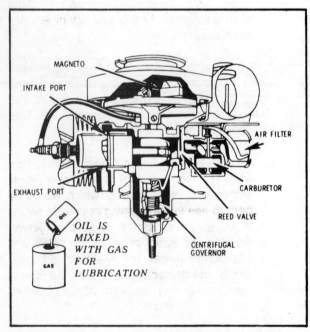

FIGURE 306 Typical 2-Cycle Engine Cross Section

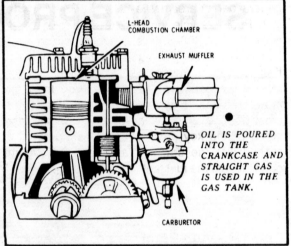

FIGURE 307 Typical-4-Cycle Engine Cross Section

Safe and Efficient Mower Operation

• This section lists helpful hints which if followed will provide safer, more efficient and trouble-free mowing.

Know Your Mower

1. The first consideration for the lawn mower owner is to be thoroughly familiar with the mower as a machine. The mower operator should know where each and every control is on the machine and the name of each general part. Fig. 305 shows a rotary lawn mower and the mower general part or area name.

2. Fuel—2-cycle or 4-cycle (2-stroke or 4-stroke). Fig. 306 shows a cross-section of a typical 2- cycle engine and some of the key parts. If the mower manufacturer's instructions tell you to mix the gasoline with oil as fuel, then the mower engine is two cycle. Fig. 307, shows a cross-section of a typical 4-cycle engine and some of the key parts. When the engine oil is poured into the engine crankcase and straight gasoline into the fuel tank, the engine is 4-cycle.

Good Mowing Practice

• After starting the engine the following mowing tips will make the job easier:

1. If grass height is unusually high cut the grass in two passes. Set the mower the highest it will go for first cut, then down to the height desired for final cut.

2. Operate the engine at the slowest speed that will do the job. Avoid long periods of high speed, maximum load operation.

3. Never operate the mower without the proper shrouds, air deflector, etc., or with the air filter removed. Running the engine without the air filter admits dirt into the engine cylinder. Even the smallest amount of dirt will wear out the engine piston rings in a matter of only a few hours.

4. Avoid running the mower across loose gravel on roads, walks or driveways. Shut the machine off before crossing.

5. Do not operate the mower in wet grass since the mower cuts this grass poorly. In addition the chances of slipping are higher and you may injure yourself.

Safety Considerations

• The next consideration is to operate the lawn mower with full recognition that it is a machine, and as a machine it represents a potential hazard if not handled properly. Accidents with lawn mowers do not just happen—they are created. The following list of safety items are recommended for accident-free mowing.

1. Walk over the area to be mowed before starting lawn mower. Remove stones, sticks, any debris, and be certain items such as garden hoses and sprinklers are not in the area.

2. Never allow children to operate the mower. Do not permit children in the general area as a safety precaution against the mower striking and throwing small stones, etc. This applies to pets as well.

3. Do the lawn mowing in daylight hours. This helps to avoid striking or running over pipes, stones and other objects which would be hard to see in the evening or twilight hours. If an object is struck, stop the mower. Remove the spark-plug ignition cable and check the mower for extent of damage. Repair the damage and be sure the mower is free of any foreign objects before attempting a restart.

4. Always set the mower cutting height with the machine turned OFF and on level and firm ground.

5. Don't start the mower until actually ready to cut. Be certain drive control is disengaged on self-propelled machines—before you start machine. Keep feet clear of blade.

6. Keep hands and feet away from operating parts such as the blades. Do not stand in front of self-propelled machines as the drive unit may engage and drive the machine into you.

7. Don't stand next to the discharge door or chute on rotary machines. Always operate the mower with discharge door closed, or with bag attached. This minimizes objects being thrown out of the chute which may hurt a bystander.

8. The engine should be started out of doors, never in an enclosed area. Lack of ventilation will permit poisonous carbon monoxide fumes to build up in the enclosed area. These fumes are deadly and can quickly overcome and kill a person in the enclosed area.

9. If the mower has a washout port always keep the port closed when mowing.

10. If for any reason the mower must be worked on, always shut the mower down and remove the spark-plug ignition cable from the plug. This prevents the engine from accidentally starting and perhaps causing a serious injury.

11. Wear closed toe shoes and long trousers while mowing. This will minimize the chance of injury due to striking loose small gravel or stones.

12. Keep a firm hold on the mower handle. Never run with the mower and never pull the mower. If you trip, the mower can run over your feet and cause a serious injury.

13. When mowing on slope, follow around the contour of the slope, across the face, never straight up or straight down. Change direction on a slope with great care least the mower be tipped over and come into contact with any part of your body.

14. Never leave the running mower unattended. A small child can be attracted to it in a matter of seconds, and can suffer grevious injury. Always shut down the mower when leaving it.

15. Always shut the mower off when adding fuel. Wait until the engine cools off before refueling. Keep flames, lit cigarettes, etc., away from the fuel while refueling. Use a strainer-equipped funnel when filling the fuel tank to avoid fuel spillage onto the engine.

16. Clean away grease and oil only when the engine is cool and use a commercial solvent or cleaner. NEVER USE GASOLINE.